Springer Theses

Recognizing Outstanding Ph.D. Research

For further volumes:
http://www.springer.com/series/8790

Aims and Scope

The series "Springer Theses" brings together a selection of the very best Ph.D. theses from around the world and across the physical sciences. Nominated and endorsed by two recognized specialists, each published volume has been selected for its scientific excellence and the high impact of its contents for the pertinent field of research. For greater accessibility to non-specialists, the published versions include an extended introduction, as well as a foreword by the student's supervisor explaining the special relevance of the work for the field. As a whole, the series will provide a valuable resource both for newcomers to the research fields described, and for other scientists seeking detailed background information on special questions. Finally, it provides an accredited documentation of the valuable contributions made by today's younger generation of scientists.

Theses are accepted into the series by invited nomination only and must fulfill all of the following criteria

- They must be written in good English.
- The topic should fall within the confines of Chemistry, Physics and related interdisciplinary fields such as Materials, Nanoscience, Chemical Engineering, Complex Systems and Biophysics.
- The work reported in the thesis must represent a significant scientific advance.
- If the thesis includes previously published material, permission to reproduce this must be gained from the respective copyright holder.
- They must have been examined and passed during the 12 months prior to nomination.
- Each thesis should include a foreword by the supervisor outlining the significance of its content.
- The theses should have a clearly defined structure including an introduction accessible to scientists not expert in that particular field.

Michael Leitner

Studying Atomic Dynamics with Coherent X-rays

Doctoral Thesis accepted by
University of Vienna, Austria

 Springer

Author
Dr. Michael Leitner
Faculty of Physics
University of Vienna
Strudlhofgasse 4
1090 Vienna
Austria
e-mail: michael.leitner@univie.ac.at

Supervisor
Dr. Bogdan Sepiol
Faculty of Physics
University of Vienna
Strudlhofgasse 4
1090 Vienna
Austria
e-mail: bogdan.sepiol@univie.ac.at

ISSN 2190-5053
ISBN 978-3-642-24120-8
DOI 10.1007/978-3-642-24121-5
Springer Heidelberg Dordrecht London New York

e-ISSN 2190-5061
e-ISBN 978-3-642-24121-5

Library of Congress Control Number: 2011938011

Cover design: eStudio Calamar, Berlin/Figueres

Printed on acid-free paper

Springer is part of Springer Science+Business Media (www.springer.com)

Supervisor's Foreword

The importance of diffusion for various processes in materials with discrete, crystalline structure or quasi-continuous like liquids and glasses is unquestionable. Diffusion takes place in solids also at moderate temperatures, although it was for a long time doubtful because of the lack of direct experimental evidence. Today, it is widely accepted that determining the temperature-driven motion of atoms in solids is more complex than figuring out the structure of solid matter. Nevertheless, it is evident that dynamical properties and especially atomic motion play a crucial role in microstructural changes occurring during preparation, processing and heat treatment of many materials, particularly in nanostructures.

Nearly all information on diffusion was derived hitherto from studies of radiotracer atoms. These studies, however, could not provide direct information about the *atomistic diffusion mechanism*. Gaining knowledge how individual atoms move and measuring their hopping rate is hence the aim of many experiments and theoretical studies. The full information about the mechanisms at work can be obtained by scattering methods being sensitive to the relevant length- and time-scales. Only few methods have satisfactory properties, such as Mößbauer spectroscopy, quasi-elastic neutron scattering and resonant scattering of synchrotron radiation. The highly desirable feat of probing the dynamics with atomic spatial and temporal resolution turns out to be extremely challenging due to several serious limitations, such as the restricted number of suitable isotopes or the limited achievable energy resolution, which necessitates very high temperatures of measurements where most matter is not even in the solid state any more.

New methods for diffusion studies were therefore greatly appreciated. Dynamic light scattering of visible light is clearly such a method and it has grown over the last three decades into a mature technique for studying dynamics in polymers, solutions of macromolecules, microemulsions, gels etc. However, it is impossible to study the atomic-scale dynamics by light scattering because atomic length scales are inaccessible and moreover solids are typically opaque for the visible light. The availability of coherent X-rays of sufficient intensity in the last decade has enabled photon correlation spectroscopy studies of diffusion. This thesis deals with the application of X-ray Photon Correlation Spectroscopy (XPCS) for answering the question how individual atoms move.

The merit of this thesis is that it could for the first time prove that atomistic dynamics is indeed measurable by XPCS using the well known model system copper-gold as a sample. While XPCS is an established method for the study of slow dynamics on length-scales of a few nanometres, it has up to now never been applied to the region of high wave-vector transfers (or large scattering angles). The scattered intensity in the diffuse regime, i.e. corresponding to atomic distances, is much lower, therefore using XPCS for this problem has always been thought impossible.

The progress achieved within this thesis is threefold: it proposes a number of systems selected for high diffuse intensity, it optimizes the photon detection and data evaluation procedures in order to obtain as much information as possible, and it establishes the theoretical models necessary for interpretating the results. These advances allowed the first successful atomic-scale XPCS experiment, which elucidated the role of preferred configurations on the atoms' jumps in a copper-gold alloy.

The copper-gold system and the few other systems treated in this thesis could still be studied with contemporary synchrotrons. The results of this thesis have already inspired a number of further investigations in different metallic and non-metallic ordered systems, as well as in solids with structural disorder such as glasses. Further work on even more elaborated systems is in progress. Moreover, more powerful sources of coherent radiation like energy recovery linacs or free electron lasers guarantee that the number of systems accessible for diffusion studies will enlarge to practically all compositions including even light-element alloys, but will also extend the upper limit of hopping rates. Thus it seems to me that this technique is the most realistic and future-oriented from the whole spectrum of experimental atomistic methods.

Vienna, July 2011   Bogdan Sepiol

Acknowledgments

In the first place I want to thank the Austrian tax-payer. As a certain wise man used to say: "If the public funds our toys, then the least that we can do is to use them for something meaningful." Being aware of the fact that I am in this respect in a very privileged position, I strived to live up to this demand.

Bogdan Sepiol was my supervisor for this thesis. Bogdan, your encyclopaedic knowledge about and understanding of physics still impresses me. Thank you for the occasions when we strayed from the topic at hand and traded ideas about what else we could possibly measure. The experimental part of this work, from the sample preparation to the furnaces and the homebrewed power supply, to which you are apparently very attached, would not have been possible without you.

Gero Vogl was until autumn 2009 the full professor in our group and is the driving force behind the application of "diffusion" aspects to fields outside of physics, in which I took part and that are not part of this thesis. These range from the spread of invasive plants to the Paleo-Indians and the dynamics of languages. Gero, it was you from whom I have learned most of the so-called soft facts about physics, or more generally science and being a scientist. You with your gifts in entertainment and guidance have—and I hope that you will continue to do—formed us hive of physicists into a group.

Lorenz-Mathias Stadler and Bastian Pfau, only your preparatory work in our group has made this effort possible. Thank you for introducing me to the field during my diploma thesis and for supporting us with the experiments. Bastian, you took part in each beamtime within the scope of this thesis, you even deserted your own beamline back in Berlin for us. Your experience with synchrotrons and your carefree (but nevertheless working) approaches towards problems with the instrumentation were very valuable. And Lorenz, now you even sacrifice your off time in order to relieve me of the writing of project proposals. This means really much to me.

Friedrich Gröstlinger has contributed manpower for the beamtimes, a really exhausting job, and it seems that I won't be able to repay him. Fritz, thank you for that, and it seems that I was even half-way successful in persuading you of the possibility of going to France by train and of the ethical inappropriateness of air travel.

Manfred Smolik, you are a really likable fellow, and I can consider myself lucky that I could share my room with you.

Philipp Haslinger, I want to thank you for suggesting about eight years ago to stay on in the lecture hall after the linear algebra course for the introductory physics course. Supposedly, it was rather amusing and what would I do all day long else, then. Today I cannot really retrace how it came to my actually deciding to starting to study physics additionally to mathematics, which resulted in solid state physics making up half of my life today, but you surely had a large part in it.

Alice, you (and recently our son Matthias) are the other half of my life. I do not thank you for your understanding on the occasions when I was only able to leave the institute so that I could still find an open shop, as this would reduce you to supporting me, while in fact you have your own life and career and are not less determined than me in pursuing it. No, I want to thank you for sharing my agitation when I recounted incidents from the office, and for the occasions when I was full of enthusiasm about a new insight and could tell you at great length, with numerous disgressions, about it, and you heard me out.

Contents

1	**Introduction**	1
	References	4
2	**Theory**	5
	2.1 Setting the Scene.	5
	2.2 The Self-Correlation Function.	6
	2.3 The Pair-Correlation Function.	13
	References	18
3	**Linking Theory to Experiments**	21
	3.1 Methods for Measuring Atomic Diffusion	21
	3.2 Theory of Scattering	22
	3.3 From Particles on a Lattice to Solid Matter	25
	3.4 Coherent and Incoherent Scattering	27
	3.5 Correlated Jumps.	28
	3.6 Theory of XPCS	29
	References	32
4	**Characteristics of Diffusion in Selected Systems**	33
	4.1 An Open System: Si_xGe_{1-x} with the Diamond Lattice	33
	4.2 A Triple Defect System: Ni-Rich B2 NiAl	36
	4.3 A Short-Range Ordered System: $Cu_{90}Au_{10}$	43
	4.4 A Model of an Amorphous System	46
	References	48
5	**Data Evaluation**.	51
	5.1 Raw Data Files	51
	5.2 Subtracting the Dark Current	53
	5.3 Histogram of the Droplet Charges	56
	5.4 Detecting Photon Events	58

5.5   Computing the Auto-Correlation Function   59
5.6   Fitting the Auto-Correlation Functions   63
References   64

6   Considerations Concerning the Experiment   65
6.1   Counting Noise   65
6.2   Optics and Contrast   68
References   74

7   Experimental Results   75
7.1 $Cu_{90}Au_{10}$ 75
7.2 $Fe_{65}Al_{35}$ 79
7.3 $Si_{89}Ge_{11}$ 81
7.4   The Metallic Glass $Zr_{65}Cu_{17.5}Ni_{10}Al_{7.5}$ 83
References   85

8   Outlook   87

Appendix   89

Curriculum Vitae   95

Chapter 1
Introduction

Solid matter is the very embodiment of permanence. In most aspects in every-day life this permanence is also a deciding factor dominating the choice of used materials: you want your can opener to cut the can without being affected, you want your bicycle to still work after being left in the rain, your house to withstand a storm, and the tungsten filament in your light bulb should survive temperatures of even thousands of degrees.[1]

This solidness on the macroscopic scale does not necessarily translate to the microscopic scale, however. The atoms oscillate about the places where they are supposed to be, and sometimes they even jump from one lattice site to the next. This perpetual motion on the atomic scale is called *dynamics*. Especially diffusion, which is the smearing-out of concentration gradients brought about by the stochastic jumps of the atoms, is a phenomenon with far-reaching consequences. It is important from the first stages of the lifespan of a product, for instance in surface hardening of tools, precipitation hardening of aluminium elements, or the doping of semiconductors, to the last stages, being responsible for corrosion or the disappearance of the doped layers due to interdiffusion. Therefore it is necessary to know about diffusion for controlling the material properties during production or for preventing their deterioration.

The rates of the atomic jumping cover many magnitudes: in the golden wedding ring you possibly have on your finger on the order of one atom jumps per second at ambient temperatures, but in metals at elevated temperatures each atom can easily jump millions of times per second. Still, because each jump happens on such short timescales,[2] at any given moment the vast majority of atoms does not jump, therefore the metal retains its macroscopic "solid" properties.

This thesis deals with the random jumping of the atoms in a solid from one stable site to the next. It will only treat the case of equilibrium dynamics, this means that there is no change of macroscopic properties with time, changes occur only on the microscopic level when the distinct atoms change place. Equivalently put, the

---

[1] Hopefully for a long time before you are forced to replace it by a compact fluorescent lamp.

[2] On the order of picoseconds, i.e., the inverse of the Debye frequency.

M. Leitner, *Studying Atomic Dynamics with Coherent X-rays*,
Springer Theses, DOI: 10.1007/978-3-642-24121-5_1,
© Springer-Verlag Berlin Heidelberg 2012

probability for a system to evolve from state A to state B is equal to the probability for evolving from B to A. Therefore applying the term diffusion to this process of random hopping is a bit misleading, because diffusion implies a spreading-out. In fact this random movement *leads* to diffusion given a concentration gradient, but it is not synonymous to it. Methods which treat interdiffusion via preparing a concentration gradient or treat tracer diffusion via preparing an isotopic gradient and measuring the smearing-out of this gradient during annealing therefore do not conform to this criterion. They directly measure diffusion but can deduce information about the random atomic movement only in an indirect way. This thesis deals in the description and direct measurement of the underlying dynamics. Unfortunately the term *diffusion* has come to be applied also for this random hopping without any spreading of gradients (as in "a particle diffuses on a lattice"), and in this sense it will also be used throughout this thesis. I beg the reader to keep this ambiguity in mind. For on overview on the fundamentals of diffusion in solid matter see the monograph of Mehrer [5], for a collection of accounts on the different aspects of diffusion and experimental methods see Heitjans and Kärger [4].

The main point of this thesis is to bring together two physical fields to solve known problems by a new method: on the one hand there is X-ray photon correlation spectroscopy (XPCS), a scattering method capable of detecting the dynamics in the sample via following the fluctuations in the scattered intensity. Until now, this method has been applied to the study of the dynamics at scattering vectors $\vec{q}$ with only very small absolute values, corresponding to comparatively large scales in real space [1, 3, 7]. On the other hand there exists a community (with strong contributions from my group, the group of Sepiol and Vogl in Vienna) devoted to the study of atomic jumps, using mainly neutron scattering and to a smaller part Mößbauer spectroscopy, see e.g. [11] for a review. Combining the existing concepts and knowledge with the non-resonant scattering method XPCS, where we have already gathered experience [6, 9], gives a very promising new tool, able to overcome the problems inherent in the existing methods (such as favourable elements or isotopes and high jump frequencies being necessary for these methods), especially in view of the imminent becoming operational of new and very powerful X-ray sources such as PETRA III and the European XFEL, both in Hamburg, and the LCLS in Stanford.

The principle of X-ray photon correlation spectroscopy is very simple: as the scattered intensity is the absolute square of the Fourier transform of the scatterer density in the sample, having disorder in the sample will result in disorder (i.e. fluctuations) in the diffuse scattering. If the atoms in the sample change their position, the scattered intensity at a given point on the detector will fluctuate over time, and XPCS essentially just records the time scale of this fluctuating as a function of the scattering vector $\vec{q}$. The crucial point for this argumentation to hold is the coherence of the incoming radiation, however. If it is not coherent, one point on the detector corresponds to a distribution of scattering vectors $\vec{q}$, and incoherent addition of their intensities smears out the fluctuations. If the incident radiation is coherent (that means a well-defined plane wave[3]), however, the equating of scattering and Fourier transforming holds, and the scattered radiation shows fluctuations, also called "speckles". The first coherent X-ray beam of sufficient intensity and therefore the first observation of X-ray speckles was reported by Sutton et al. [10]. Coherent optical light, however, has been available since the invention of the laser in the 1960s, and the analogous method, which is most often called Dynamic Light Scattering, has matured to a standard method of characterization of soft matter or emulsions. Many overviews of the method exist, see, e.g., [2].

Obtaining coherent radiation is in principle not difficult, all one has to do is to take an incoherent source and cut out a sufficiently small volume in phase space by using slits (for transversal coherence) and monochromators (for longitudinal coherence). This is the way it is done nowadays in X-ray physics, as a synchrotron is an incoherent source. With such an approach one obviously trades intensity for coherence. This is a fundamental problem, because increasing the detection efficiency is no option, XPCS is already at the single-photon level. In the future inherently coherent X-ray sources, free-electron lasers in the hard X-ray regime, will become available, just as lasers are in the optical regime. This will open up a vast range of possibilities for coherent X-ray physics, enabling the study of processes inaccessible up to now.

What XPCS detects is the stochastic evolution of the disorder in the sample. Applied to the case of atomic dynamics, it therefore detects chemical diffusion, that means the process that governs how a given occupation of the lattice sites by atoms of different kinds (or by atoms and holes) evolves to another occupation. This is a difference to incoherent methods such as Mößbauer spectroscopy, which measure tracer diffusion, i.e. they label the atoms (by radioactivity or by imprinting a phase onto the nuclei) and measure how a given labelled atom diffuses.

In the frame of this thesis the first successful applications of XPCS to the problem of studying atomic diffusion were performed. Apart from documenting the results and physical insights obtained from these experiments another important point for me was to devote some space both to the fundamental theoretical aspects of atomic diffusion and to considerations concerning the experimental side, such that it would ideally be possible for any solid state physicist to start with this subject using this thesis alone, without having to reinvent all the small details a second time.

This is the structure of the thesis: first the concepts for describing stochastic motion on a lattice are introduced and results relevant to this thesis are derived, and these theoretical concepts are linked to the case of atomic diffusion and to the results obtained by an XPCS experiment. Then a number of systems exemplary for distinct aspects of atomic diffusion are presented. In the next chapter the evaluation of the raw data obtained with an XPCS experiment is described in some detail, followed by a chapter about optimizing the experimental set-up. Then comes a chapter about the experimental results obtained on the aforementioned systems so far, and finally an outlook is given. Everything presented in this thesis is original, apart from

---

[3] Note that for the appearance of speckles only a fixed phase relation in space and time is necessary, but in order to be able to use the Fourier transform I require plane waves, which will be fulfilled in a good approximation within the small illuminated sample volume.

Chap. 3 (which for the most part can be found in textbooks) and the section on the self-correlation function in Chap. 2. The derivation of the temporal evolution of the pair-correlation function under the constraint of short-range order given in Sect. 2.3 is also original, it re-derives the result already given by Sinha and Ross [8] for the case of quasi-elastic neutron scattering in a fashion that is more easily understandable to physicists used to diffusion on a lattice.

References

1. S. Brauer, G.B. Stephenson, M. Sutton, R. Brüning, E. Dufresne, S.G.J. Mochrie, G. Grübel, J. Als-Nielsen, D.L. Abernathy, X-Ray intensity fluctuation spectroscopy observation of critical dynamics in Fe_3Al. Phys. Rev. Lett. **74**, 2010 (1995)
2. W. Brown (ed.), *Dynamic Light Scattering: The Method and Some Applications* (Clarendon Press, Oxford, 1993)
3. S.B. Dierker, R. Pindak, R.M. Fleming, I.K. Robinson, L. Berman, X-Ray photon correlation spectroscopy study of Brownian motion of gold colloids in glycerol. Phys. Rev. Lett. **75**, 449 (1995)
4. P. Heitjans, J. Kärger (eds.), *Diffusion in Condensed Matter*, 2nd edn. (Springer, Berlin, 2005)
5. H. Mehrer, *Diffusion in solids: Fundamentals Methods Materials Diffusion-Controlled Processes* (Springer, Berlin, 2007)
6. B. Pfau, L.-M. Stadler, B. Sepiol, R. Weinkamer, J. Kantelhardt, F. Zontone, G. Vogl, Coarsening dynamics in elastically anisotropic alloys. Phys. Rev. B **73**, 180101 (2006)
7. O.G. Shpyrko et al., Direct measurement of antiferromagnetic domain fluctuations. Nature (London) **447**, 68 (2007)
8. S.K. Sinha, D.K. Ross, Self-consistent density response function method for dynamics of light interstitials in crystals. Physica B **149**, 51 (1988)
9. L.-M. Stadler, B. Sepiol, R. Weinkamer, M. Hartmann, P. Fratzl, J. Kantelhardt, F. Zontone, G. Grübel, G. Vogl, Long-term correlations distinguish coarsening mechanisms in alloys. Phys. Rev. B **68**, 180101 (2003)
10. M. Sutton, S.G.J. Mochrie, T. Greytak, S.E. Nagler, L.E. Berman, G.A. Held, G.B. Stephenson, Observation of speckle by diffraction with coherent X-rays. Nature (London) **352**, 608 (1991)
11. G. Vogl, B. Sepiol, The elementary diffusion step in metals studied by the interference of gamma-rays, X-Rays and Neutrons. In: P. Heitjans, J. Kärger (eds.), *Diffusion in Condensed Matter*, 2nd edn. (Springer, Berlin, 2005) pp. 65–91

Chapter 2
Theory

In this chapter I will give the fundamental concepts to describe diffusion on the atomic level, with special consideration to the case of diffusion on a lattice. For reasons of conceptual simplicity this chapter will stay abstract, I will work out the connection to the actual problem at hand—atomic diffusion in condensed matter—in Chap. 3.

2.1 Setting the Scene

The system to be described consists of particles diffusing in infinite space. The particles behave equally but are distinguishable. The theoretical tools for the problem at hand, i.e. to describe the stochastic motion of the particles, were given by van Hove [12]. These are:

- The *self-correlation function* $G_s(\Delta \vec{x}, \Delta t)$ gives the probability to find a *given* particle at time $t + \Delta t$ at the position $\vec{x} + \Delta \vec{x}$ given that it (the *same* particle) was at time t at position $\vec{x}$.
- The *pair-correlation function* $G(\Delta \vec{x}, \Delta t)$ gives the probability to find *any* particle at time $t + \Delta t$ at the position $\vec{x} + \Delta x$ given that *any* particle was at time t at position $\vec{x}$.

The definition given above is the classical case of van Hove's quantum-mechanical theory. This is justified by the fact that first for the systems of interest in this thesis the spatial uncertainty of the particles is given by thermal excitations and not quantum effects (e.g. tunneling) and second that the scattering of X-rays on diffusing atoms can be considered truly elastic due to the X-rays' high energy. The formulation with time differences instead of the correlations between two absolute times implies that the system is in equilibrium. Obviously the functions given above do not contain all the information of the dynamic process. One could continue and consider correlation functions of higher order, e.g. the probability of a particle being at a given time and place if it was at time t_1 at place $\vec{x}_1$ and at time t_2 at place $\vec{x}_2$, but for diffusion modelled as a Markov process the description by two-point correlations suffices.

M. Leitner, *Studying Atomic Dynamics with Coherent X-rays*,
Springer Theses, DOI: 10.1007/978-3-642-24121-5_2,
© Springer-Verlag Berlin Heidelberg 2012

From now on I will restrict my attention to particles diffusing on a lattice. I assume the lattice to be three-dimensional, as this covers all cases treated later in this thesis, but this is just for convenience, the reader is invited to picture a lattice of arbitrary finite dimensionality,[1] everything given here generalizes. Let the lattice be composed of Λ sublattices. The translation vectors of the fundamental lattice be $\vec{a}_1, \vec{a}_2$, and $\vec{a}_3$, being linearly independent, but not necessarily orthogonal. I use this basis set for spanning $\mathbb{R}^3$. The distinct sites in the sublattice have the coordinate vectors r_λ for $1 \leq \lambda \leq \Lambda$. Finally I define the vectors spanning reciprocal space $\vec{b}_1, \vec{b}_2$, and $\vec{b}_3$ such that $\vec{a}_i \cdot \vec{b}_j = 2\pi \delta_{i,j}$. This set of vectors can easily be constructed:

$$\vec{b}_1 = 2\pi \frac{\vec{a}_2 \times \vec{a}_3}{\vec{a}_1 \cdot (\vec{a}_2 \times \vec{a}_3)}, \tag{2.1.1}$$

$\vec{b}_2$ and $\vec{b}_3$ follow by cyclic permutation. Given a vector $\vec{x}$ relative to the Cartesian unity vectors $\vec{e}_i$ I will write x for its coordinate vector relative to the translation vectors of the lattice $\vec{a}_i$, analogously with a reciprocal vector $\vec{q}$ and its coordinate vector q relative to the reciprocal lattice vectors $\vec{b}_i$. This has the property that the valid positions of the particles are given by $x + r_\lambda$ for $x \in \mathbb{Z}^3$ and $1 \leq \lambda \leq \Lambda$. Also note that $q \cdot x = \vec{q} \cdot \vec{x}$ due to the definition of the reciprocal lattice vectors.

Having this definitions out of the way, I now move on to the description of the dynamics, the correlation functions.

2.2 The Self-Correlation Function

As stated above, van Hove's self-correlation function $G_s(\Delta x, \Delta t)$ gives the conditional probability for a given particle to be at time $t + \Delta t$ at position $x + \Delta x$ under the condition that this particle was at time t at position x. The reason for treating the self-correlation function is first that it is a rather intuitive way of describing dynamics and second that there are methods which (more or less) directly measure it (see Sect. 3.4). These methods realize the measurement of the probability via the actual displacements of a vast number of atoms. I will now deduce the temporal evolution of this probability density. For past approaches to this problem see [2, 6, 7, 9, 10].

I consider particles diffusing on a lattice. As its name already tells, for the self-correlation function the movement of a particle with respect to itself alone is of relevance. Therefore it suffices to consider (the probability distribution of) the positions of one particle over time. In reality particles can interact, so actually the temporal evolution of the tagged particle's position is influenced by the configuration of its surrounding. As I describe the state of the system only by the position of the one

---

[1] Diffusion on a surface would be a physically relevant case of diffusion on a lower-dimensional lattice.

tagged particle, this fact can lead to a non-Markovian behaviour of the system (earlier states of the system can influence the hidden variables, i.e. the configuration of the neighbourhood, influencing in turn the further evolution). The simplification which makes the problem tractable is to postulate Markovian behaviour, i.e. that the probability distribution of the states of the system at some later time are only a function of the state of the system now.

The temporal evolution of the probability density is therefore defined by specifying the transition rates between the sites on the lattice. I write $(\mathbf{K}(\Delta x))_{\mu,\lambda}$ for the transition rate of the particle from sublattice λ in the cell x to sublattice μ in the cell $x + \Delta x$. Put another way, the entry in row μ, column λ of the matrix $\mathbf{K}(\Delta x)$ multiplied by an infinitesimal amount of time is the probability for a particle on the sublattice λ to jump onto the site μ of the cell displaced by Δx within this amount of time. For mass conservation I put the overall leaving rate from the sublattice λ into $(\mathbf{K}(0))_{\lambda,\lambda}$ but counted negatively:

$$\sum_{(\Delta x,\mu)\neq(0,\lambda)} (\mathbf{K}(\Delta x))_{\mu,\lambda} = -(\mathbf{K}(0))_{\lambda,\lambda}. \qquad (2.2.1)$$

I require detailed balance, this means that in equilibrium there should be no net flux between two states of the system:

$$(\mathbf{K}(\Delta x))_{\mu,\lambda} p_\lambda = (\mathbf{K}(-\Delta x))_{\lambda,\mu} p_\mu \qquad (2.2.2)$$

where p_λ is the equilibrium probability for a particle to reside on sublattice λ.

My goal is to compute the temporal evolution of probabilities, so I introduce an ensemble of systems (i.e. an ensemble of particles). This ensemble is completely specified by $(c(x,t))_\lambda$, the concentration (i.e. the ratio) at time t of the particles in the ensemble which reside in the cell x on the sublattice λ. With above definition of $\mathbf{K}$ the temporal derivative of the concentration c can now be written as

$$(\dot{c}(.,t))_\lambda = \sum_\mu (\mathbf{K}(.))_{\lambda,\mu} * (c(.,t))_\mu, \qquad (2.2.3)$$

where the symbol $*$ denotes convolution in space. This now explains where $\mathbf{K}$ got its symbol: it is the matrix-valued diffusion kernel.

Just as such equations are customarily solved I apply the element-wise spatial Fourier transformation, that means $\mathcal{F}((c(.,t))_\lambda) = (\hat{c}(.,t))_{\hat{\lambda}}$ analogously for $\mathbf{K}$:

$$(\dot{\hat{c}}(q,t))_\lambda = \sum_\mu (\hat{\mathbf{K}}(q))_{\lambda,\mu} \cdot (\hat{c}(q,t))_\mu, \qquad (2.2.4)$$

or put more elegantly

$$\dot{\hat{c}}(q,t) = \hat{\mathbf{K}}(q) \cdot \hat{c}(q,t), \qquad (2.2.5)$$

understood as matrix multiplication. An equation like that is one of the first problems encountered in the analysis of ordinary differential equations. Defining exponentiation for matrices via the series expansion of the scalar-valued exponential function, the solution to this ordinary differential can be immediately given:

$$\hat{c}(\boldsymbol{q}, t) = e^{\hat{\mathbf{K}}(\boldsymbol{q})t} \cdot \hat{c}(\boldsymbol{q}, 0). \tag{2.2.6}$$

Remembering that the vector notation of the concentration is just a shorthand for a scalar-valued concentration of the form $\sum_\lambda \left(c(., t)\right)_\lambda * \delta(. - \boldsymbol{r}_\lambda)$, its Fourier transform is therefore $\sum_\lambda \left(\hat{c}(., t)\right)_\lambda \exp(-i\boldsymbol{q}\boldsymbol{r}_\lambda)$.

Let now $f_\mu^\lambda(\Delta\boldsymbol{x}, \Delta t)$ be the probability distribution for finding a particle at time Δt on the site μ of cell $\Delta\boldsymbol{x}$ if it was at time 0 at site λ of cell $\mathbf{o}$. The spatial Fourier transform of this function is $\left(e^{\hat{\mathbf{K}}(\boldsymbol{q})\Delta t}\right)_{\mu,\lambda}$ (use Eq. 2.2.6 with an initial condition c equal to 1 at site $\mathbf{o}$ and sublattice λ and take entry μ of the result). Just considering the particles from sublattice λ would give for the self-correlation function $\sum_\mu f_\mu^\lambda(x - \boldsymbol{r}_\mu + \boldsymbol{r}_\lambda, \Delta t)$. Taking into account the particles starting from all sublattices with their respective weights p_λ gives

$$G_s(\Delta\boldsymbol{x}, \Delta t) = \sum_\mu \sum_\lambda f_\mu^\lambda(\Delta\boldsymbol{x} - \boldsymbol{r}_\mu + \boldsymbol{r}_\lambda, \Delta t) p_\lambda, \tag{2.2.7}$$

and in the Fourier domain

$$I_s(\boldsymbol{q}, \Delta t) := \mathcal{F}\left(G_s(., \Delta t)\right)(\boldsymbol{q}) = \sum_\mu e^{-i\boldsymbol{q}\boldsymbol{r}_\mu} \sum_\lambda \left(e^{\hat{\mathbf{K}}(\boldsymbol{q})\Delta t}\right)_{\mu,\lambda} p_\lambda e^{i\boldsymbol{q}\boldsymbol{r}_\lambda}. \tag{2.2.8}$$

For reasons that will become clear in Sect. 3.4, I_s goes under the name incoherent intermediate scattering function.

To put Eq. 2.2.8 more elegantly, I first define the $1 \times \Lambda$-matrix $\mathbf{E} = (e^{-i\boldsymbol{q}\boldsymbol{r}_1} \ldots e^{-i\boldsymbol{q}\boldsymbol{r}_\Lambda})$, the $\Lambda \times \Lambda$-diagonal matrix P with the entries $p_1 \ldots p_\Lambda$ in the diagonal, and the Hermitized diffusion kernel in reciprocal space

$$\mathbf{K}'(\boldsymbol{q}) := \sqrt{\mathsf{P}^{-1}} \hat{\mathbf{K}}(\boldsymbol{q}) \sqrt{\mathsf{P}}. \tag{2.2.9}$$

Because each component of $\mathbf{K}(\Delta\boldsymbol{x})$ is real, taking the component-wise complex conjugation of its Fourier transform is equivalent to inverting the independent variable:

$$\hat{\mathbf{K}}(-\boldsymbol{q}) = \overline{\hat{\mathbf{K}}(\boldsymbol{q})}. \tag{2.2.10}$$

To show that $\mathbf{K}'(\boldsymbol{q})$ is actually Hermitian I first restate Eq. 2.2.2:

$$\mathbf{K}(\Delta\boldsymbol{x})\mathsf{P} = \left(\mathbf{K}(-\Delta\boldsymbol{x})\mathsf{P}\right)^\mathsf{T} = \mathsf{P}\mathbf{K}^\mathsf{T}(-\Delta\boldsymbol{x}), \tag{2.2.11}$$

which naturally also holds for its Fourier transform

$$\hat{\mathbf{K}}(q)\mathbf{P} = \mathbf{P}\hat{\mathbf{K}}^{\mathrm{T}}(-q). \tag{2.2.12}$$

Multiplying this equality from both sides by $\sqrt{\mathbf{P}^{-1}}$ and using Eq. 2.2.10 leads to

$$\sqrt{\mathbf{P}^{-1}}\hat{\mathbf{K}}(q)\sqrt{\mathbf{P}} = \sqrt{\mathbf{P}}\hat{\mathbf{K}}^{\mathrm{T}}(-q)\sqrt{\mathbf{P}^{-1}} = \sqrt{\mathbf{P}}\hat{\mathbf{K}}^{*}(q)\sqrt{\mathbf{P}^{-1}}, \tag{2.2.13}$$

where $(\ldots)^{*}$ denotes the adjoint matrix, thereby proving the claim.

With these definitions Eq. 2.2.8 reads

$$I_s(q, \Delta t) = \mathbf{E}(q)\exp\left(\hat{\mathbf{K}}(q)\Delta t\right)\mathbf{P}\mathbf{E}^{*}(q) = \mathbf{E}(q)\exp\left(\sqrt{\mathbf{P}}\mathbf{K}'(q)\sqrt{\mathbf{P}^{-1}}\Delta t\right)\mathbf{P}\mathbf{E}^{*}(q)$$
$$= \mathbf{E}(q)\sqrt{\mathbf{P}}\exp\left(\mathbf{K}'(q)\Delta t\right)\sqrt{\mathbf{P}}\mathbf{E}^{*}(q). \tag{2.2.14}$$

From

$$\overline{I_s(q, \Delta t)} = I_s(q, \Delta t)^{*} = \left(\mathbf{E}(q)\sqrt{\mathbf{P}}\exp\left(\mathbf{K}'(q)\Delta t\right)\sqrt{\mathbf{P}}\mathbf{E}^{*}(q)\right)^{*}$$
$$= \mathbf{E}^{**}(q)\sqrt{\mathbf{P}^{*}}\exp\left(\mathbf{K}'^{*}(q)\Delta t\right)\sqrt{\mathbf{P}^{*}}\mathbf{E}^{*}(q)$$
$$= \mathbf{E}(q)\sqrt{\mathbf{P}}\exp\left(\mathbf{K}'(q)\Delta t\right)\sqrt{\mathbf{P}}\mathbf{E}^{*}(q)$$
$$= I_s(q, \Delta t), \tag{2.2.15}$$

where the first equality followed trivially from considering I_s an 1×1-matrix and the following equalities from the Hermitianness of $\mathbf{K}'$ and the rules for matrix transposition, it follows that I_s is real. Using this fact, Eq. 2.2.10 and the definition of $\mathbf{E}(q)$

$$I_s(q, \Delta t) = \overline{I_s(q, \Delta t)} = \mathbf{E}(-q)\sqrt{\mathbf{P}}\exp\left(\mathbf{K}'(-q)\Delta t\right)\sqrt{\mathbf{P}}\mathbf{E}^{*}(-q) = I_s(-q, \Delta t), \tag{2.2.16}$$

so I_s is in fact even and real-valued. Therefore also G_s, being the back-transform of an real-valued even function, is even (and real-valued).

This is at first glance surprising, as the lattice's being composed of sublattices will in the general case destroy the inversion symmetry of the underlying Bravais lattice. The key point in above derivation, however, was the invocation of detailed balance. This principle just says that the same number of atoms hop from site A to site B as from site B to site A, so even if the fluxes exiting site A have no inversion symmetry, the other sites make up for that imbalance, leading to an even correlation function.

It is instructive to write I_s in yet another way. Diagonalizing $\mathbf{K}'$, i.e. writing

$$\mathbf{K}'(q) = \mathbf{V}(q)\mathbf{D}(q)\mathbf{V}^{*}(q), \tag{2.2.17}$$

with $\mathbf{V}(q)$ a unitary matrix and $\mathbf{D}(q)$ a diagonal matrix with real (because $\mathbf{K}'$ is Hermitian) non-positive (see Sect. A.1) diagonal entries, Eq. 2.2.14 reads

$$\begin{aligned}
I_s(\boldsymbol{q}, \Delta t) &= \mathbf{E}(\boldsymbol{q})\sqrt{\mathbf{P}} \exp\left(\mathbf{V}(\boldsymbol{q})\mathbf{D}(\boldsymbol{q})\mathbf{V}^*(\boldsymbol{q})\Delta t\right)\sqrt{\mathbf{P}}\mathbf{E}^*(\boldsymbol{q}) \\
&= \mathbf{E}(\boldsymbol{q})\sqrt{\mathbf{P}}\mathbf{V}(\boldsymbol{q}) \exp\left(\mathbf{D}(\boldsymbol{q})\Delta t\right)\mathbf{V}^*(\boldsymbol{q})\sqrt{\mathbf{P}}\mathbf{E}^*(\boldsymbol{q}) \\
&= \sum_\lambda e^{(\mathbf{D}(\boldsymbol{q}))_{\lambda,\lambda}\Delta t} \left| \sum_\mu e^{-i\boldsymbol{q}\boldsymbol{r}_\mu} \sqrt{p_\mu}\left(\mathbf{V}(\boldsymbol{q})\right)_{\mu,\lambda} \right|^2.
\end{aligned} \tag{2.2.18}$$

$I_s(\boldsymbol{q}, \Delta t)$ for a fixed $\boldsymbol{q}$ is therefore a sum of Λ (possibly degenerate) exponential decays, where the respective decay times are given by the inverse of the diagonal entries in $\mathbf{D}(\boldsymbol{q})$ and the respective weights are a function of the occupation probabilities of the sublattices p_λ, the geometry within the unit cell $\boldsymbol{r}_\lambda$ in relation to $\boldsymbol{q}$, and the jump frequencies between the various sites.

I want to point out an analogy of the present problem to another one most solid state physicists are probably more familiar with: phonon dispersion. In a crystal composed of Λ sublattices there are Λ phonon states for a given wave-vector $\boldsymbol{q}$, one acoustic and $\Lambda - 1$ optical phonons. The eigenvalues of $\mathbf{K}'(\boldsymbol{q})$ (which are the diagonal entries of $\mathbf{D}(\boldsymbol{q})$) behave similarly: for small $\boldsymbol{q}$ they can be divided into one value describing the decay of long-range correlations and $\Lambda - 1$ values describing the fluxes between the sublattices. Considerations along the lines of the proof in Sect. A.1 show that the appearance of an additional eigenvalue equal to zero at a $\boldsymbol{q}$ equal to a reciprocal lattice vector (apart from the "acoustic" eigenvalue) is equivalent to the lattice's decomposing into two (or more) systems of sublattices, so that there is no flux from sites in one system to sites in the other (in the phonon analogy this would correspond to the artificial example of two interleaved lattices which do not interact, leading to an optical phonon branch behaving like an additional acoustic branch). A non-trivial case is the interstitialcy mechanism of diffusion in the diamond lattice, see Sect. 4.1. In this case there is no flux between the two sublattices, so for a $\boldsymbol{q}$ equal to a reciprocal lattice vector the intermediate incoherent scattering function does not decay in time.

I want to treat now Eq. 2.2.18 in the limit of small $\boldsymbol{q}$ for the non-degenerate case, i.e. where the lattice does not decompose. Using Eq. 2.2.1 it follows that the diagonal vector of $\sqrt{\mathbf{P}}$, in the following denoted $\sqrt{\mathbf{p}}$, is the "acoustic" eigenvector of $\mathbf{K}'(\mathbf{o})$ corresponding to the eigenvalue 0. As $\sqrt{\mathbf{P}}\mathbf{E}^*(\boldsymbol{q})$ converges to $\sqrt{\mathbf{p}}$ for $\boldsymbol{q} \to \mathbf{o}$, the weight of the "optical" decays in Eq. 2.2.18 vanishes, leaving only the "acoustic" decay, the eigenvalue of which goes to 0. For computing the behaviour at small $\boldsymbol{q}$ of this eigenvalue, in the following denoted $d(\boldsymbol{q})$ and defined by the equation

$$d(\boldsymbol{q})\mathbf{v}(\boldsymbol{q}) = \mathbf{K}'(\boldsymbol{q})\mathbf{v}(\boldsymbol{q}), \tag{2.2.19}$$

I write the relevant quantities as power series in $\boldsymbol{q}$:

$$\begin{aligned}
d(\boldsymbol{q}) &= d^0 + d^1(\boldsymbol{q}) + d^2(\boldsymbol{q}) + O(\boldsymbol{q}^3), \\
\mathbf{v}(\boldsymbol{q}) &= \mathbf{v}^0 + \mathbf{v}^1(\boldsymbol{q}) + \mathbf{v}^2(\boldsymbol{q}) + O(\boldsymbol{q}^3), \\
\mathbf{K}'(\boldsymbol{q}) &= \sqrt{\mathbf{P}^{-1}}\left(\mathbf{K}^0 + \mathbf{K}^1(\boldsymbol{q}) + \mathbf{K}^2(\boldsymbol{q})\right)\sqrt{\mathbf{P}} + O(\boldsymbol{q}^3).
\end{aligned} \tag{2.2.20}$$

Here quantities with 1 in the exponent are linear functions of $\boldsymbol{q}$ and quantities with 2 are bilinear functions. Expanding in a power series is valid because in the non-degenerate case $d(q)$ has a multiplicity of 1 everywhere around $q = o$ and is therefore an analytical function of the coefficients of the characteristic polynomial of $\mathbf{K}'(q)$. By construction d^0 is 0 and $\mathbf{v}^0$ is $\sqrt{\mathbf{p}}$, but also d^1 is 0 because $\mathbf{K}'(q)$ and therefore also its eigenvalues are even functions in q. This leads to the necessity for the linear terms in q on the right-hand side of Eq. 2.2.19 to cancel for all q, therefore

$$\mathbf{K}^1(q)\mathbf{p} + \mathbf{K}^0\sqrt{\mathbf{P}}\mathbf{v}^1 = 0. \tag{2.2.21}$$

$\mathbf{K}^0$ is not invertible, so $\sqrt{\mathbf{P}}\mathbf{v}^1$ is given by

$$\sqrt{\mathbf{P}}\mathbf{v}^1 = -(\mathbf{K}^0)^{-1}\mathbf{K}^1(q)\mathbf{p} + s\mathbf{p}. \tag{2.2.22}$$

Here $(\mathbf{K}^0)^{-1}$ denotes the Moore-Penrose pseudoinverse of $\mathbf{K}^0$ and $s\mathbf{p}$ spans the kernel of $\mathbf{K}^0$ due to Eqs. 2.2.1 and 2.2.2 and the fact that the rank of the kernel is 1.

The quantity of interest $d(q)$ follows then as

$$\begin{aligned}
d(q) &= \mathbf{v}(q)^{\mathrm{T}}d_a(q)\mathbf{v}(q) = \mathbf{v}(q)^{\mathrm{T}}\mathbf{K}'(q)\mathbf{K}(q) \\
&= \mathbf{v}(q)^{\mathrm{T}}\sqrt{\mathbf{P}^{-1}}\left(\mathbf{K}^2\sqrt{\mathbf{P}}\mathbf{v}^0 + \mathbf{K}^1\sqrt{\mathbf{P}}\mathbf{v}^1 + \mathbf{K}^0\sqrt{\mathbf{P}}\mathbf{v}^2\right) + O(q^3) \\
&= \mathbf{e}\left(\mathbf{K}^2\sqrt{\mathbf{P}}\mathbf{v}^0 + \mathbf{K}^1\sqrt{\mathbf{P}}\mathbf{v}^1\right) + O(q^3) \\
&= \mathbf{e}\left(\mathbf{K}^2(q) - \mathbf{K}^1(q)(\mathbf{K}^0)^{-1}\mathbf{K}^1(q)\right)\mathbf{p} + O(q^3)
\end{aligned} \tag{2.2.23}$$

with $\mathbf{e} = (1, \ldots, 1)$ and observing the cancellation of various terms. The relevant matrices are explicitly given by

$$\begin{aligned}
\mathbf{K}^0 &= \sum_{\Delta x} \mathbf{K}(\Delta x), \\
\mathbf{K}^1(q) &= -\mathrm{i}\sum_{\Delta x} (q\,\Delta x)\mathbf{K}(\Delta x), \\
\mathbf{K}^2(q) &= -\sum_{\Delta x} \frac{(q\,\Delta x)^2}{2}\mathbf{K}(\Delta x),
\end{aligned} \tag{2.2.24}$$

so Eq. 2.2.23 is a non-negative (see Sect. A.1) quadratic form for small q

$$d(q) = q^{\mathrm{T}}\mathbf{D}q + O(q^3), \tag{2.2.25}$$

and the intermediate incoherent scattering function Eq. 2.2.18 reads for small q

$$I_s(q, \Delta t) = e^{-q^{\mathrm{T}}\mathbf{D}q\,\Delta t}. \tag{2.2.26}$$

The description of diffusion in macroscopic terms is given by Fick's laws. In Fick's first law the diffusion tensor $\mathbf{D}$ is defined via the phenomenological linear relation between concentration gradient and mass flux

$$j = -\mathbf{D}\nabla c. \tag{2.2.27}$$

Invoking mass conservation leads to Fick's second law

$$\dot{c} = \nabla \mathbf{D} \nabla c, \tag{2.2.28}$$

or in reciprocal space

$$\dot{\hat{c}}(\boldsymbol{q}, t) = -\boldsymbol{q}\mathbf{D}q\hat{c}(\boldsymbol{q}, t). \tag{2.2.29}$$

Solving this equation with a delta distribution as initial condition gives

$$\hat{c}(\boldsymbol{q}, t) = e^{-\boldsymbol{q}^{\mathrm{T}}\mathbf{D}qt}. \tag{2.2.30}$$

Therefore the quadratic form $\mathbf{D}$ describing the behaviour of the intermediate incoherent scattering function in Eq. 2.2.26 at small $\boldsymbol{q}$ is nothing else than the macroscopic diffusion tensor. It is given by the macroscopic limit of the self-correlation function which is experimentally mainly determined from the spreading of a small amount of radioactive tracer atoms, so it is customarily called the tracer diffusion tensor.

In the degenerate case, where the lattice decomposes into mutually disconnected sets of sublattices, the problem can be solved on each set alone. Note that this can give different quadratic forms $\mathbf{D}$ for the distinct sets of sublattice. Therefore the non-degeneracy assumption in the derivation of Eq. 2.2.26 is not just for convenience, in fact in the general case the macroscopic description by Fick's laws is not valid. For the above-mentioned case of interstitialcy diffusion in the diamond lattice both sublattices behave equally at small $\boldsymbol{q}$, so the partial intermediate incoherent scattering functions can be merged and the phenomenological macroscopic behaviour is recovered.

In the special case where the particles sit on a Bravais lattice and therefore all sites are equivalent, the diffusion kernel $K(\Delta x)$ is scalar-valued, has inversion symmetry and $\sum_{\Delta x} K(\Delta x) = 0$. Defining

$$\Gamma_{\mathrm{inc}}(\boldsymbol{q}) = -\hat{K}(\boldsymbol{q}) = -\mathcal{F}(K)(\boldsymbol{q}) = -\sum_{\Delta x} K(\Delta x)\cos(\boldsymbol{q}\Delta x)$$

$$= \sum_{\Delta x} K(\Delta x)\big(1 - \cos(\boldsymbol{q}\Delta x)\big), \tag{2.2.31}$$

Eq. 2.2.18 has the concise form

$$I_s(\boldsymbol{q}, \Delta t) = e^{-\Gamma_{\mathrm{inc}}(\boldsymbol{q})\Delta t}. \tag{2.2.32}$$

Γ_{inc} is called the incoherent linewidth, as quasi-elastic methods measure it as a line broadening (see Sect. 3.1). In the Bravais case Eq. 2.2.23 and equivalently Eq. 2.2.31 simplify to

$$d(\boldsymbol{q}) = \Gamma_{\mathrm{inc}}(\boldsymbol{q}) = \sum_{\Delta x} K(\Delta x)\frac{(\boldsymbol{q}\Delta x)^2}{2} \tag{2.2.33}$$

for small q. Additionally invoking cubic symmetry leads to the ellipsoid described by the quadratic form **D** becoming a sphere, therefore diffusion becomes isotropic, completely specified by the scalar-valued tracer diffusion constant D with

$$D = \sum_{\Delta x} K(\Delta x) \frac{|\Delta x|^2}{6}. \tag{2.2.34}$$

This equation is called the Einstein relation, where the additional factor 3 in the denominator compared to Eq. 2.2.33 is due to the value of $(q \Delta x)^2$ averaged over all directions being

$$\left\langle (q \Delta x)^2 \right\rangle = \frac{|q|^2 |\Delta x|^2}{3}. \tag{2.2.35}$$

It is worth reflecting on the approximations inherent in this section's considerations. First, in actual metallic systems in most cases diffusion does not happen by spontaneous, unprovoked hopping, rather it is the result of the migration of a vacancy. This fact leads to correlations between hopping events. These correlations can, however, be satisfactorily incorporated into the model in the framework of the so-called encounter model (see Sect. 3.5). The second issue, the influence of the particle's surroundings, was already addressed at the beginning of this section. With interacting particles the jump probabilities are not a strict function of the initial and target sublattices, but they vary with the surroundings. This leads to the fact that there is not one single well-defined decay time per sublattice, so the decay gets "stretched" due to averaging over the distinct surroundings, corresponding to different exponential decays. Still, in most experimental studies this effect is not drastic, and the data can be fitted by one exponential per sublattice.

2.3 The Pair-Correlation Function

This section gives an analytic theory of the pair-correlation function in the case of short-range order on the lattice. The simplification introduced in Sect. 2.2, i.e. describing the systems in the ensemble only by the position of the tagged particle and accounting for the different surroundings only through an average "effective" surrounding, does not work here, as the very essence of the pair correlation function lies in the correlation with *other* particles. Therefore the temporal evolution of the positions of all the particles in the system has to be described in a unified approach, explicitly treating the correlations. Nevertheless, in order to obtain analytic results, some approximations have to be made, namely the high-temperature limit, i.e. to treat correlations due to the energetics as first-order perturbations. A further assumption is that the Hamiltonian is given via pairwise interactions. Considering several sublattices would only obfuscate the ideas presented here, so I assume the particles to sit on sites of a Bravais lattice. The derivation presented here, which uses classical transition state theory [13], leads to the same results as the one given by Sinha and Ross [11] set in a general framework of lattice dynamics. In the meantime is has been published as Leitner and Vogl [8]. For a still quicker, although less fundamental treatment of this problem see the last paragraphs of this section.

A state of the system is described by the occupation function σ, that is, $\sigma(x) = 1$ if the site x is occupied by a particle and $\sigma(x) = 0$ if not. The most general Hamiltonian for pair potentials is given by

$$H(\sigma) = V_0 + V_1 \sum_x \sigma(x) + \sum_{x,y} V(x - y)\sigma(x)\sigma(y). \qquad (2.3.1)$$

In the following only the differences between energies of states in the canonical ensemble will be required, so the expression

$$H'(\sigma) = \sum_{x,y} V(x - y)\sigma(x)\sigma(y) \qquad (2.3.2)$$

can be used without loss of generality. I write

$$\Delta E(x; \Delta x, \sigma) = H'(\sigma_2) - H'(\sigma_1) \qquad (2.3.3)$$

for the difference in energy between a state σ_1 with

$$\sigma_1(y) = \begin{cases} 1 & y = x \\ 0 & y = x + \Delta x \\ \sigma(y) & \text{else} \end{cases} \qquad (2.3.4)$$

and a state σ_2 with

$$\sigma_2(y) = \begin{cases} 0 & y = x \\ 1 & y = x + \Delta x \\ \sigma(y) & \text{else.} \end{cases} \qquad (2.3.5)$$

Described in words, $\Delta E(x; \Delta x, \sigma)$ is the energy gained (or lost) when moving a particle from x to $x + \Delta x$ in the environment specified by σ. I write $E_s(x, x + \Delta x; \sigma)$ for the energy of the saddle point in the energy landscape on the path from x to Δx, relative to the average of the initial and final energy $H'(\sigma_1)$ and $H'(\sigma_2)$. Equivalently stated, the energy necessary to invest for raising a particle on its way from x to Δx onto the saddle point is given by $E_s(x, x + \Delta x; \sigma) + \Delta E(x; \Delta x, \sigma)/2$, for moving it back it is $E_s(x, x + \Delta x; \sigma) - \Delta E(x; \Delta x, \sigma)/2$. These concepts will become more clear by means of an example in Sect. 4.3.

There are two equivalent possible choices for the fundamental dynamic process: either the particles hop into empty sites (and do not hop if the prospective target site is occupied) or the occupancy of two sites is exchanged, i.e. if exactly one of the two is occupied, after the exchange the other is occupied, if either both are occupied or both are unoccupied, nothing changes. I use the latter concept, as it is symmetric under the operation $\sigma \to 1 - \sigma$. The rate of exchanges of the occupances of x and $x + \Delta x$ can then be written

$$\omega = \nu_{\Delta x} e^{-\frac{\Delta E(x; \Delta x, \sigma)}{2k_B T}} e^{-\frac{E_s(x, x+\Delta x; \sigma)}{k_B T}}$$

$$= \nu_{\Delta x} e^{-\frac{E_s(\Delta x)}{k_B T}} e^{-\frac{\Delta E(x; \Delta x, \sigma)}{2k_B T}} e^{-\frac{\Delta E_s(x, x+\Delta x; \sigma)}{k_B T}}, \tag{2.3.6}$$

where $E_s(\Delta x)$ is the mean saddle point energy for a jump along Δx with $\Delta E_s(x, x + \Delta x; \sigma)$ being the variations around this mean value and $\nu_{\Delta x}$ the attempt frequency for such a jump [13].

I will now treat the temporal evolution of the system, where the system is initially in the state σ. Obviously a given site x can either be occupied or unoccupied, so $\sigma(x)$ is either 0 or 1. It will turn out that the equation describing the evolution is in first order linear in σ, therefore the same relationship holds also for the expected value, that is the average value over an ensemble of systems. The reader is invited to choose the most convenient setting, either a concrete state and transition probabilities or expected values and their temporal evolution.

$$\dot{\sigma}(x) = \sum_{\Delta x} \left(\sigma(x + \Delta x)(1 - \sigma(x)) e^{-\frac{\Delta E(x; \Delta x, \sigma)}{2k_B T}} \right.$$

$$\left. -\sigma(x)(1 - \sigma(x + \Delta x)) e^{-\frac{\Delta E(x; \Delta x, \sigma)}{2k_B T}} \right) \nu_{\Delta x} e^{-\frac{E_s(x, x+\Delta x; \sigma)}{k_B T}}$$

$$= \sum_{\Delta x} \nu_{\Delta x} e^{-\frac{E_s(\Delta x)}{k_B T}} e^{-\frac{\Delta E_s(x, x+\Delta x; \sigma)}{k_B T}}$$

$$\times \left(\sigma(x + \Delta x)(1 - \sigma(x)) e^{-\frac{\Delta E(x; \Delta x, \sigma)}{2k_B T}} \right.$$

$$\left. -\sigma(x)(1 - \sigma(x + \Delta x)) e^{-\frac{\Delta E(x; \Delta x, \sigma)}{2k_B T}} \right)$$

$$= \sum_{\Delta x} \tilde{\nu}_{\Delta x} \left((\sigma(x + \Delta x) - \sigma(x)) \left(1 - \frac{\Delta E_s(x, x + \Delta x; \sigma)}{k_B T} \right) \right.$$

$$+ \frac{\Delta E(x; \Delta x, \sigma)}{2k_B T} (\sigma(x) + \sigma(x + \Delta x)$$

$$\left. -2\sigma(x)\sigma(x + \Delta x)) + O\left((E/k_B T)^2 \right) \right) \tag{2.3.7}$$

Here E is a measure for the typical energy variations, in both the saddle point and the stable positions, i.e. $\Delta E(x; \Delta x, \sigma) = O(E)$ and $\Delta E_s(x, x + \Delta x; \sigma) = O(E)$. $\tilde{\nu}_{\Delta x} = \nu_{\Delta x} \exp(-E_s(\Delta x)/k_B T)$ is the raw jump frequency neglecting the influence of energy and correlations.

The system is assumed to be only short-range ordered, this means that

$$\langle \sigma(x)\sigma(y) \rangle - \langle \sigma(x) \rangle \langle \sigma(y) \rangle = O(E/k_B T) \tag{2.3.8}$$

for $x \neq y$. As both $\Delta E(x; \Delta x, \sigma)$ and $\Delta E_s(x, x + \Delta x; \sigma)$ are linear functionals with respect to σ which depend neither on $\sigma(x)$ nor on $\sigma(x + \Delta x)$, it follows that

$$\langle \Delta E(x; \Delta x, \sigma)\sigma(x) \rangle - \langle \Delta E(x; \Delta x, \sigma) \rangle \langle \sigma(x) \rangle = O\left((E/k_B T)^2\right), \qquad (2.3.9)$$

analogously for similar quantities. Applying this to Eq. 2.3.7 and cancelling (noting that $\langle \sigma(x) \rangle$ is equal to the concentration of particles c) shows that

$$\dot{\sigma}(x) = \sum_{\Delta x} \tilde{v}_{\Delta x} \left(\sigma(x + \Delta x) - \sigma(x) + \frac{\Delta E(x; \Delta x, \sigma)}{k_B T} c(1 - c) \right) \qquad (2.3.10)$$

in first order approximation, in particular the influence of the configuration on the energetics of the saddle point vanishes, only the energies of the initial and the final state matter.

Going back to Eq. 2.3.2, $\Delta E(x; \Delta x, \sigma)$ is explicitly given by

$$\Delta E(x; \Delta x, \sigma) = \sum_y V(x + \Delta x - y)\sigma(y) - \sum_y V(x - y)\sigma(y)$$

$$= \sum_z V(z)\big(\sigma(x + \Delta x - z) - \sigma(x - z)\big). \qquad (2.3.11)$$

Defining the amplitude $A = \mathcal{F}(\sigma)$ and using basic results about the Fourier transform of convolutions, the transform in x of above equation reads

$$\mathcal{F}\big(\Delta E(.; \Delta x, \sigma)\big)(q) = \hat{V}(q)A(q)(e^{iq\Delta x} - 1), \qquad (2.3.12)$$

where $\hat{V}$ is the transform of the pair potential V. Using this result the Fourier transform of Eq. 2.3.10 can be given as

$$\dot{A}(q) = \sum_{\Delta x} \tilde{v}_{\Delta x} \left(A(q)e^{iq\Delta x} - A(q) + \frac{\hat{V}(q)A(q)(e^{iq\Delta x} - 1)}{k_B T} c(1 - c) \right)$$

$$= A(q) \sum_{\Delta x} \tilde{v}_{\Delta x} \big(\cos(q\Delta x) - 1\big) \left(1 + \frac{\hat{V}(q)c(1 - c)}{k_B T} \right). \qquad (2.3.13)$$

In this section's nomenclatura Γ_{inc} now reads (cp. Eq. 2.2.31)

$$\Gamma_{\text{inc}}(q) = \sum_{\Delta x} \tilde{v}_{\Delta x} \big(1 - \cos(q\Delta x)\big), \qquad (2.3.14)$$

but this time the relevant quantity is the coherent linewidth

$$\Gamma_{\text{coh}}(q) = \Gamma_{\text{inc}}(q) \left(1 + \frac{\hat{V}(q)c(1 - c)}{k_B T} \right), \qquad (2.3.15)$$

giving

$$\langle A(\boldsymbol{q}, t)\rangle = A(\boldsymbol{q}, 0)e^{-\Gamma_{\text{coh}}(\boldsymbol{q})t}. \tag{2.3.16}$$

Writing the time dependence of σ explicitly, the pair-correlation function $G(\Delta\boldsymbol{x}, \Delta t)$ is defined by

$$G(\Delta\boldsymbol{x}, \Delta t) = \langle\sigma(., .)\sigma(. + \Delta\boldsymbol{x}, . + \Delta t)\rangle. \tag{2.3.17}$$

Note that

$$G(\Delta\boldsymbol{x}, \Delta t) = \langle\sigma(., .)\sigma(. + \Delta\boldsymbol{x}, . + \Delta t)\rangle = \langle\sigma(. - \Delta\boldsymbol{x}, . - \Delta t)\sigma(., .)\rangle$$
$$= \langle\sigma(., .)\sigma(. - \Delta\boldsymbol{x}, . - \Delta t)\rangle = G(-\Delta\boldsymbol{x}, -\Delta t). \tag{2.3.18}$$

Due to time-inversion symmetry G is even in time, using this fact and above result it is also even in space.

Again using the interplay of Fourier transforming and convoluting the coherent intermediate scattering function is given by

$$I(\boldsymbol{q}, \Delta t) := \mathcal{F}(G(., \Delta t))(\boldsymbol{q}) = \langle A(\boldsymbol{q}, .)\hat{A}(\boldsymbol{q}, . + \Delta t)\rangle = \langle A(\boldsymbol{q}+, .)\hat{A}(\boldsymbol{q}, .)e^{-\Gamma_{\text{coh}}(\boldsymbol{q})\Delta t}\rangle$$
$$= I_{\text{SRO}}(\boldsymbol{q})e^{-\Gamma_{\text{coh}}(\boldsymbol{q})\Delta t}. \tag{2.3.19}$$

$I_{\text{SRO}}(\boldsymbol{q})$ is the expected value of the intensity, the squared modulus of the amplitude, due to short-range order, for a given $\boldsymbol{q}$.

In the framework of the approximations invoked here the intensity can be directly related to the potential via the Clapp-Moss-relations [3], see Sect. A.2:

$$I_{\text{SRO}}(\boldsymbol{q}) = \frac{1}{\left(1 + \frac{\hat{V}(\boldsymbol{q})c(1-c)}{kT}\right)}, \tag{2.3.20}$$

therefore

$$\Gamma_{\text{coh}}(\boldsymbol{q}) = \frac{\Gamma_{\text{inc}}(\boldsymbol{q})}{I_{\text{SRO}}(\boldsymbol{q})}. \tag{2.3.21}$$

The intensity $I_{\text{SRO}}(\boldsymbol{q})$ is measured in Laue units, where one Laue unit is $Nc(1 - c)$ with N the number of lattice sites (this is just the value of the configurational diffuse scattering of a random alloy). In particular it follows that the coherent linewidth is equal to the incoherent linewidth for vanishing interactions, and $I_s = I$.

Just as with the intermediate incoherent scattering function in Sect. 2.2 also here the behaviour of $\Gamma_{\text{coh}}(\boldsymbol{q})$ for small $\boldsymbol{q}$ is given by a quadratic form corresponding to a diffusion tensor $\mathbf{D}$. In this case, however, it describes the decay of chemical fluctuations in the macroscopic limit, I will therefore call it the chemical diffusion tensor (or chemical diffusion constant in the cubic case). An analogon of the Einstein relation also holds here, by Eq. 2.3.21 the diffusion constant is just the value of the tracer diffusion constant in Eq. 2.2.34 divided by $I_{\text{SRO}}(\mathbf{o})$.

The fact that the relaxation of fluctuations (i.e. the decay of the coherent intermediate scattering function) becomes slower than the value given by Chudley and Elloitt

[2] for positions in reciprocal space with high intensity has been known qualitatively under the name de Gennes-narrowing [5] from studies of liquids and colloidal glasses [1, 4]. It is not difficult to understand: a high $I_{SRO}(q)$ means that the particles prefer to build local arrangements corresponding to a high Fourier component at q. The reason can only be that such arrangements are energetically favoured compared to other arrangements, therefore it takes more energy to break such arrangements up, making them longer-lived. However, the fact that the decay is in the first approximation still a single exponential given by the very simple relation (2.3.21), being only a function of the static energetics, is not so obvious. The simulations in Sect. 4.3 elucidate what happens when the approximations invoked here break down.

I want also to sketch another, less fundamental way of deriving Eq. 2.3.21. I write the pair-correlation function as the sum of the self-correlation function and the distinct-correlation function

$$G(\Delta x, \Delta t) = G_s(\Delta x, \Delta t) + G_d(\Delta x, \Delta t), \tag{2.3.22}$$

equivalently in reciprocal space

$$I(q, \Delta t) = I_s(q, \Delta t) + I_d(q, \Delta t). \tag{2.3.23}$$

The behaviour of $I_s(q, \Delta t)$ was derived in Sect. 2.2, it decays with the rate $\Gamma_{inc}(q)$. The assumption of equilibrium leads to time-inversion symmetry, therefore Eq. 2.2.32 can be generalized to negative time differences

$$I_s(q, \Delta t) = e^{-\Gamma_{inc}(q)|\Delta t|}. \tag{2.3.24}$$

$G_d(\Delta x, \Delta t)$ and therefore $I_d(q, \Delta t)$ can be seen as the reaction of the surrounding particles to the occupation of site o. This reaction happens via diffusion and will therefore vary smoothly in time, just as the heat conduction equation smoothes out singularities in the initial or boundary data. $I_d(q, \Delta t)$ obviously also has time-inversion symmetry, so with it being smooth everywhere it has a vanishing temporal derivative at $\Delta t = 0$. Therefore

$$-\Gamma_{coh}(q) I_{SRO}(q) = \frac{d}{d\Delta t} I(q, \Delta t)\Big|_{\Delta t=0^+} = \frac{d}{d\Delta t} I_s(q, \Delta t)\Big|_{\Delta t=0^+} = -\Gamma_{inc}(q), \tag{2.3.25}$$

and Eq. 2.3.21 follows.

References

1. C. Caronna, Y. Chushkin, A. Madsen, A. Cupane, Dynamics of nanoparticles in a supercooled liquid. Phys. Rev. Lett. **100**, 055702 (2008)
2. C.T. Chudley, R.J. Elliott, Neutron scattering from a liquid on a jump diffusion model. Proc. Phys. Soc. Lond. **77**, 353 (1961)
3. P.C. Clapp, S.C. Moss, Correlation functions of disordered binary alloys. I. Phys. Rev. **142**, 418 (1966)

4. B.A. Dasannacharya, K.R. Rao, Neutron scattering from liquid argon. Phys. Rev. **137**, A417 (1965)
5. P.G. de Gennes, Liquid dynamics and inelastic scattering of neutrons. Physica **25**, 825 (1959)
6. M.A. Krivoglaz, The effect of diffusion on the scattering of neutrons and photons by crystal imperfections and on the Mössbauer effect. Sov. Phys. JETP **13**, 1273 (1961)
7. R. Kutner, I. Sosnowska, Thermal neutron scattering from a hydrogen-metal system in terms of a general multi-sublattice jump diffusion model–I: theory. J. Phys. Chem. Solids **38**, 741 (1977)
8. M. Leitner, G. Vogl, Quasi-elastic scattering under short-range order: the linear regime and beyond. J. Phys. Condens. Matter **23**, 254206 (2011)
9. O.G. Randl, B. Sepiol, G. Vogl, R. Feldwisch, K. Schroeder, Quasielastic Mössbauer spectroscopy and quasielastic neutron scattering from non-Bravais lattices with differently occupied sublattices. Phys. Rev. B **49**, 8768 (1994)
10. J.M. Rowe, K. Sköld, H.E. Flotow, J.J. Rush, Quasielastic neutron scattering by hydrogen in the α and β phases of vanadium hydride. J. Phys. Chem. Solids **32**, 41 (1971)
11. S.K. Sinha, D.K. Ross, Self-consistent density response function method for dynamics of light interstitials in crystals. Physica B **149**, 51 (1988)
12. L. van Hove, Correlations in space and time and born approximation scattering in systems of interacting particles. Phys. Rev. **95**, 249 (1954)
13. G.H. Vineyard, Frequency factors and isotope effects in solid state rate processes. J. Phys. Chem. Solids **3**, 121 (1957)

Chapter 3
Linking Theory to Experiments

In Chap. 2 the theory of the self- and the pair-correlation function for particles diffusing on a lattice was given. This chapter will give the connection to the scientific problem at hand, studying atomic diffusion in crystalline matter.

3.1 Methods for Measuring Atomic Diffusion

The most capable experimental techniques for this kind of research are scattering techniques. The major ones of these are (listed chronologically):

- *Quasi-elastic neutron scattering (QENS)*. This method analyzes the energy transferred from the sample to the neutron as a function of q. Due to the relation $E = \hbar\omega$, this constitutes an experimental determination of the temporal Fourier transform of the intermediate scattering function, the so-called dynamic structure factor $S(q, \omega)$. As the neutrons are scattered by the nuclei, they are sensitive to the distinct isotopes. Using this fact and fine-tuning the isotope composition of the sample often either the coherent or the incoherent scattering can be made to dominate (see Sect. 3.4), which corresponds to measuring the self- or the pair-correlation function. QENS is in principle a very versatile method, but first, not all elements are suited to this method, and second, the available flux is rather low, implying long measuring times. Furthermore, the technical improvement of neutron sources has happened only at a slow pace during the last decades, so this situation cannot be expected to improve much in the near future.
- *Mößbauer spectroscopy*. This method exploits the fact that certain nuclei have very narrow energetical states and that in a solid there is a non-vanishing probability for nuclear transitions to happen without interference of thermal vibrations. If an atom diffuses while undergoing a nuclear transition this leads to a broadening of the linewidth due to the emitted wave train apparently decomposing into several sub-trains with distorted phase relation at the observer. Probing the line shape as a function of q by employing the Doppler shift when moving the source relative

M. Leitner, *Studying Atomic Dynamics with Coherent X-rays*,
Springer Theses, DOI: 10.1007/978-3-642-24121-5_3,
© Springer-Verlag Berlin Heidelberg 2012

to the sample again leads to $S(q, \omega)$, just as with QENS, but here the scattering is purely incoherent, also the accessible q lie on a sphere due to the fixed energy of the nuclear transition. By far the most suited isotope for Mößbauer spectroscopy is ^{57}Fe, but this essentially limits its application to systems containing iron. An advantage Mößbauer spectroscopy has over the other methods mentioned here is that it is a tabletop technique, feasible in a small laboratory.

- *Nuclear resonant scattering (NRS)*. This is in fact simply Mößbauer spectroscopy in the time domain, it therefore in principle directly measures the incoherent intermediate scattering function $I_s(q, t)$. Instead of a radioactive source it uses pulsed synchrotron radiation, thanks to which measurements are much faster and can even be done on a single atomic layer, but obviously it needs access to a synchrotron.

- *X-ray photon correlation spectroscopy (XPCS)*. This is the method this thesis deals with. In short, it correlates the temporal variation of the intensity scattered from coherent synchrotron radiation at the sample. Contrary to Mößbauer spectroscopy, nuclear resonant scattering and most implementations of QENS this yields the coherent intermediate scattering function $I(q, t)$. It is in principle not limited to certain isotopes or elements, and it has a great potential for the future due to the recent or imminent launch of new X-ray sources.

XPCS works directly in the time domain: it measures how fast the scattered intensity fluctuates. Accessible times are in principle not limited, in practice the stability of the beam allows measurements on the scale of tens of minutes or even an hour. QENS and Mößbauer spectroscopy work in the energy domain, however. Slow processes lead to narrow lines, so the limited experimental resolution dictates processes on the order of nanoseconds or faster. Also NRS is limited to this range by the natural lifetime of the excited state.[1]

Doing a scattering experiment essentially amounts to performing the spatial Fourier transform of the scatterer density (electrons or nuclei, depending on the method). The intermediate scattering function is therefore a natural way for describing the results of experiments, as it describes the processes in reciprocal space.

3.2 Theory of Scattering

This section gives the fundamentals of scattering at a classical static system in the kinematical approximation for electro-magnetic radiation and particles, see e.g. Als-Nielsen and McMorrow [1] for a thorough treatment. An illustration of the process is given in Fig. 3.1.

The electric field of a monochromatic plane electro-magnetic wave with angular frequency ω and wave-vector $\vec{k}$ is given in complex notation by

$$\vec{E}_i(\vec{x}, t) = \vec{E}_0 e^{i(\vec{k}\vec{x} - \omega t)}. \tag{3.2.1}$$

---

[1] Quasi-elastic methods are therefore ideally suited for the measurement of phonons.

Fig. 3.1 Illustration of the quantities used for describing the scattering process. dV is the differential scattering volume positioned at $\vec{x}_0$, $\vec{k}$ and $\vec{k}'$ are the wave-vectors of the incident and outgoing radiation, respectively, $\vec{q} = \vec{k}' - \vec{k}$ is the wave-vector transfer

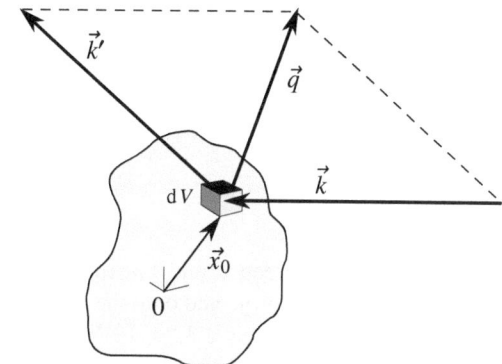

This oscillating field exerts a force on a charged particle with mass m and charge q, located at position $\vec{x}_0$,

$$\vec{F}(t) = \vec{E}_i(\vec{x}_0, t)q = \vec{E}_0 q e^{i(\vec{k}\vec{x}_0 - \omega t)}. \tag{3.2.2}$$

If the particle can be considered as free, i.e. if its eigenfrequency ω_0 is much smaller than the frequency of the incident radiation ω, this leads to a displacement of the particle from its equilibrium position

$$\vec{d}(t) = -\frac{\vec{E}_0 q}{m\omega^2} e^{i(\vec{k}\vec{x}_0 - \omega t)} \tag{3.2.3}$$

and therefore to a dipole moment

$$\vec{p}(t) = q\vec{d}(t). \tag{3.2.4}$$

In Hertz' theory the electric field at the position $\vec{x}$ far away from such an oscillating dipole at $\vec{x}_0$ is given by an outgoing spherical wave

$$\vec{E}_s(\vec{x}, t) = \frac{\omega^2}{\epsilon_0 c^2} \frac{\left((\vec{x} - \vec{x}_0) \times \vec{p}(t) \times (\vec{x} - \vec{x}_0)\right)}{|\vec{x} - \vec{x}_0|^2} G(\vec{x} - \vec{x}_0), \tag{3.2.5}$$

where

$$G(\vec{x}) = \frac{e^{ik|\vec{x}|}}{4\pi |\vec{x}|} \tag{3.2.6}$$

and $k = |\vec{k}|$. Computing the elastic far-field scattering into the direction $\vec{k}'$, that is for large $|\vec{x}|$ and $\vec{x} \parallel \vec{k}'$ with $|\vec{k}'| = |\vec{k}|$, this approximates to

$$\vec{E}_s(\vec{x}, t) = \frac{q^2}{4\pi \epsilon_0 m c^2} \frac{\vec{k}' \times \vec{E}_0 \times \vec{k}'}{k^2} \frac{e^{i(\vec{k}\vec{x} - \omega t)}}{|\vec{x}|} e^{-i(\vec{k}' - \vec{k})\vec{x}_0}. \tag{3.2.7}$$

This can immediately be generalized to an arbitrary number of scatterers, described by their number density $n(\vec{x}_0)$. Defining $\vec{q} = \vec{k}' - \vec{k}$ and $r_0 = q^2/4\pi\epsilon_0 mc^2$, the magnitude of the field is essentially the Fourier transform with respect to $\vec{q}$ of the scatterer density

$$E_s(\vec{x}) = r_0 E_0 \sin(\phi)\frac{1}{|\vec{x}|}\left(\mathcal{F}(n)\right)(\vec{q}) \tag{3.2.8}$$

with ϕ the angle between $\vec{k}'$ and $\vec{E}_0$; the direction of the field is the polarization of the incident radiation projected onto the normal plane of the exiting radiation. When scattering at electrons $r_0 = 2.82 \times 10^{-15}$m is called the Thomson scattering length.

The equivalent problem for scattering of an incident flux of particles of mass m, described by a wave function

$$\psi_i(\vec{x}) = \psi_0 e^{i\vec{k}\vec{x}}, \tag{3.2.9}$$

at a particle which interacts with the incoming particles via a potential $\frac{2\pi\hbar^2}{m}b\delta(\vec{x} - \vec{x}_0)$, with b the scattering length, is solving the time-independent Schrödinger equation

$$\left(-\frac{\hbar}{2m}\Delta + \frac{2\pi\hbar^2}{m}b\delta(\vec{x} - \vec{x}_0)\right)\psi(\vec{x}) = E\psi(\vec{x}) = \frac{\hbar^2\vec{k}^2}{2m}\psi(\vec{x}) \tag{3.2.10}$$

or

$$\left(\Delta + \vec{k}^2\right)\psi(\vec{x}) = 4\pi b\delta(\vec{x} - \vec{x}_0)\psi(\vec{x}). \tag{3.2.11}$$

Using the result that the fundamental solution for the Helmholtz operator

$$\left(-\Delta - \vec{k}^2\right)G(\vec{x}) = \delta(\vec{x}) \tag{3.2.12}$$

is an outgoing spherical wave

$$G(\vec{x}) = \frac{e^{ik|\vec{x}|}}{4\pi|\vec{x}|}, \tag{3.2.13}$$

the solution to Eq. 3.2.11 is given by a perturbation to the incoming wave

$$\psi(\vec{x}) = \psi_i(\vec{x}) - \psi_i(\vec{x}_0)4\pi bG(\vec{x} - \vec{x}_0). \tag{3.2.14}$$

In analogy to Eq. 3.2.7 the scattered part of the wave-function for scattering at one particle can be written

$$\psi_s(\vec{x}) = -b\psi_0\frac{e^{i\vec{k}\vec{x}}}{|\vec{x}|}e^{-i(\vec{k}'-\vec{k})\vec{x}_0}, \tag{3.2.15}$$

and for scattering at a system of particles with number density $n(\vec{x}_0)$ it is given by the Fourier transformation of n

$$\psi_s(\vec{x}) = -b\psi_0 \frac{e^{i\vec{k}\vec{x}}}{|\vec{x}|} \big(\mathcal{F}(n)\big)(\vec{q}). \tag{3.2.16}$$

As for both photons and particles the probability density is given by the absolute square of the electric field or wave-function, respectively, both Eqs. 3.2.8 and 3.2.16 lead to the number of photons/particles scattered into a given direction $\vec{k}'$ being given by the Fourier transform of the scatterer density

$$\left(\frac{d\sigma}{d\Omega}\right)(\vec{k}') = \sigma \left|\big(\mathcal{F}(n)\big)(\vec{k}' - \vec{k})\right|^2, \tag{3.2.17}$$

where the factor σ takes in the physics of scattering. From now on this factor (and the influence of polarization) will be disregarded[2] and the intensity as a function of the wave-vector transfer be defined as squared modulus of the Fourier transform of the scatterer density

$$I(\vec{q}) = \left|\big(\mathcal{F}(n)\big)(\vec{q})\right|^2. \tag{3.2.18}$$

Note that the number density of scatterers $n(\vec{x}_0)$ is equivalent to the description via the spin operator σ in Sect. 2.3.

Here the scattered field was treated as a perturbation of the incident field. This is called the first Born approximation or kinematical scattering. The physical situation corresponds to the thin-sample limit, i.e. the path of the radiation through the material is so short that the probability for a particle to be scattered multiple times (in the particle view) is negligible. For the scattering of X-rays, this is normally fulfilled, as the absorption cross section is orders of magnitude larger than the scattering cross section, necessitating thin samples (from the point of view of scattering) in order to have any photons exiting the sample. For other probes, e.g. neutrons, electrons, or resonant γ-quanta, this is not the case and multiple-scattering effects can be appreciable.

3.3 From Particles on a Lattice to Solid Matter

Two points need clarification in order to link experiments on real physical systems to the results of Chap. 2, pertaining to the very simple, abstract system of particles on a lattice: First, in real crystals there is not one kind of particle on a partly empty lattice, but elements (with possibly different isotopes) and vacancies or interstitials,

---

[2] Note, however, that at the ESRF both the incoming radiation is polarized in the horizontal plane and scattering is mostly done in horizontal geometry. For small-angle scattering the effect is negligible, but with a scattering angle of 45° the scattered radiation is reduced by a factor of two.

and second, possibly the atoms do not sit exactly on the positions described by the lattice (due to disorder and elastic interactions), and even the lattice itself can have defects like dislocations.

For the first point a number of cases have to be considered: In a sample consisting of only one atomic species an incoherent method sensitive to this element will obviously measure the incoherent intermediate scattering function of this species. Self-interstitials will not be visible at all (as their number is always very low), and vacancies are also not directly visible, only through their effects: the atoms' diffusivity scales with the vacancies' number, and they lead to correlated jumps (see Sect. 3.5). A coherent method, however, will *only* see the vacancies or interstitials (whichever is the dominant defect), that means, the situation corresponds to the one in Sect. 2.3 with the particles being the vacancies (or interstitials). This is because a coherent method computes the Fourier transform of the scatterer density, and the constant background of the filled lattice only affects the (unmeasurable) Fourier component for $q = 0$. The number of these defects is unfortunately always very low, ruling out this kind of experiment with today's sources.

If the sample is a solid solution of two or more elements, with incoherent methods it is in principle possible to measure the incoherent intermediate scattering functions of each constituent separately, either by using different incident radiation in the case of Mößbauer spectroscopy or nuclear resonant scattering, or by preparing isotopically different, but chemically identical samples for QENS. For a coherent method and a sample consisting of two elements, it is of no consequence to which element the status 'occupied' and to which the status 'unoccupied' in the setting of Sect. 2.3 is assigned, as the difference again is only in the Fourier component for $q = 0$. The case of the sample consisting of more elements is not within the scope of Sect. 2.3, in this case it would be necessary to treat the interactions and correlations between each two constituents separately.

If finally the sample is an ordered alloy, say an A-rich intermetallic with AB-order, where the surplus A-atoms form structural antisites, incoherent and coherent methods measure very different things: for an incoherent method sensitive to the A-atoms the generic case of Sect. 2.2 applies, i.e. diffusion on distinct sublattices with high weight of the A-sublattice and low weight of the A-atoms on the B-sublattice. For a coherent method, however, the A-sublattice would be completely invisible, as it is fully ordered (neglecting thermal defects) and therefore does not contribute to the diffuse intensity. This case would therefore correspond to Sect. 2.3, where the lattice is the B-sublattice and the particles are the structural A-antisites on the B-sublattice.

The second point in the list requiring clarification was the influence of lattice distortions and defects. In contrast to macroscopic methods such as radioactive tracer experiments, the enhanced diffusivity in the vicinity of defects (dislocations, twins, anti-phase boundaries in ordered alloys, or grain boundaries in the case of polycrystalline samples) normally does not influence the results. Tracer experiments measure the average squared displacement, which can be dominated by the effect of defects. Atomistic methods measure how fast correlations on atomic length scales decay on average, and it therefore does not matter if a very small part of the atoms (the ones in the vicinity of the defects) have displacements on the order of thousands of the atomic length scale. Such a situation would give in the intermediate scattering function two decays with very different timescales, where the fast component has a weight on the order of the volume fraction of the defect, rendering it invisible. Therefore atomistic methods intrinsically measure the equilibrium bulk diffusivity, making the question of sample preparation much less critical.

Concerning the effect of lattice distortions: these generally happen in the case of disorder, i.e. if there are sublattices which are not exclusively occupied by one element only. This breaks the symmetry of the underlying lattice, and due to elastic interactions and relaxation the atoms will be displaced in relation to their ideal geometric positions. For incoherent methods this results in the positions of the sublattices r_λ in Eq. 2.2.8 becoming a distribution, thereby smoothing the fluctuations of I_s at high q. Coherent scattering, however, is affected qualitatively by atomic displacements: Take an A–B solid solution where the two elements have different sizes. The neighbours of, say, a B-atom are displaced outwards from their average positions, the neighbours of an A-atom are displaced inwards. Working out the scattering in a first approximation, this means that a B-atom not only brings with it its distinct electron density distribution at its position, but it also induces dipoles of electron density at the positions of its neighbours, as their electron densities are moved outwards relative to their mean position. This can be taken into consideration via the atomic form factors, leading to a contribution to the diffuse intensity called displacement scattering. Therefore $I_{SRO}(q)$ in Eq. 2.3.19, which is defined as the scattering due to disorder, i.e. from point-like particles on an ideal lattice, is not directly accessible in scattering experiments. For a review of the treatments of the connection between the configuration of an alloy and its scattered radiation see Schönfeld [3].

3.4 Coherent and Incoherent Scattering

In Sect. 3.2 it was stated that the scattered intensity is the squared modulus of the Fourier transform of the scattering length density of the sample. Picture now a system of particles at the fixed positions R_n with the scattering lengths b_n (for simplicity assumed as real). The scattered intensity for a given q is

$$I(q) = \sum_{n_1,n_2} b_{n_1} b_{n_2} e^{i(R_{n_1}-R_{n_2})q}. \tag{3.4.1}$$

Assuming the scattering lengths b_n to be independent and identically distributed random variables the expected value of the intensity is

$$\langle I(q)\rangle = \sum_{n_1,n_2} \langle b_{n_1} b_{n_2}\rangle e^{i(R_{n_1}-R_{n_2})q} = \sum_{n_1} \langle b_{n_1}^2\rangle + \sum_{n_1 \neq n_2} \langle b_{n_1}\rangle \langle b_{n_2}\rangle e^{i(R_{n_1}-R_{n_2})q}$$

$$= \sum_{n_1} b_{inc}^2 + \sum_{n_1,n_2} b_{coh}^2 e^{i(R_{n_1}-R_{n_2})q}, $$

$$\tag{3.4.2}$$

where b_{inc} is termed the incoherent scattering length

$$b_{\mathrm{inc}} = \sqrt{\langle b_n^2 \rangle - \langle b_n \rangle^2} \qquad (3.4.3)$$

and b_{coh} the coherent scattering length

$$b_{\mathrm{coh}} = \langle b_n \rangle. \qquad (3.4.4)$$

Especially for neutron scattering, taking the expected value in Eq. 3.4.2 is experimentally inadvertently realized by the averaging of the detector over a range of q, therefore the detected intensity *looks as if* it was made up of a part scattered coherently at the particles (showing the correlations in the positions) with a scattering length per particle b_{coh} and of a part scattered incoherently (no angular variations) with a scattering length per particle b_{inc}. The derivation presented here shows that the scattering itself is obviously coherent, but that the deviations of the actual scattering lengths from the mean value average out of the cross terms, leading to disproportionally higher self terms and apparent incoherent scattering. This is very relevant for neutron scattering, as elements can consist of different isotopes or have different spin states, both effects lead to different scattering lengths for atoms which are chemically identical and therefore randomly distributed. Consequently the correlation function probed by QENS is a sum of a coherent and incoherent part (corresponding to pair- and self-correlation function, respectively), with the weights depending on the isotopic composition.

X-rays, however, are not scattered at the nuclei, but at the electrons. In this case there is a one-to-one correspondence between the scattering length density and the chemical configuration, and XPCS is therefore an entirely coherent method.

3.5 Correlated Jumps

In Chap. 2 the temporal evolution of the sample was assumed to be described by a Markov process. Specifically the successive jumps of a particle were assumed to be independent. In the case of solutes hopping from one interstitial site to the next, as it is the case with small atoms like hydrogen, this is a valid assumption. In the overwhelming number of metallic systems where diffusion is mediated by vacancies, however, this does not hold any more. The reason is that the equilibrium concentration of vacancies is always very small, therefore after one atom has moved into a vacancy, there is now a vacancy behind it, leading to a jump probability higher than on average (and with a tendency to reverse the jump) and thereby breaking the Markovian assumption.

In the limit of a vanishing vacancy concentration there is a solution to this problem, the so-called encounter model [2]. As the timescale of the successive jumps of a particle effected by one vacancy becomes separated from the timescale between the encounters of a particle with different vacancies, it becomes possible to treat these two stages separately. First the probabilities for the effective displacements after one encounter are calculated, this can be done by numerical solution of analytical expressions with arbitrary precision [4]. Apart from the encounters where the effective displacement is zero, the fraction of which is approximately given by the inverse of the coordination number, a few percent of the encounters lead to displacement outside of the nearest-neighbour shell. These effective displacement probabilities are then used with the theory of Chap. 2.

A high degree of order can be another reason for correlated jumps. In an ordered alloy the vacancy has to choose its way in compliance with the requirement of keeping the degree of order. This will be treated in greater detail by means of an example in Sect. 4.2.

3.6 Theory of XPCS

In this section I will work out how one measures the coherent intermediate scattering function (and thereby the pair-correlation function) in an XPCS experiment. For a more extensive treatment see Sutton [5] and the references therein.

The intensity-intensity auto-correlation function is defined by

$$G^{(2)}(\boldsymbol{q}, \Delta t) = \langle I(\boldsymbol{q}, .)I(\boldsymbol{q}, . + \Delta t)\rangle, \tag{3.6.1}$$

its normalized version reads

$$g^{(2)}(\boldsymbol{q}, \Delta t) = \frac{\langle I(\boldsymbol{q}, .)I(\boldsymbol{q}, . + \Delta t)\rangle}{\langle I(\boldsymbol{q}, .)\rangle^2}. \tag{3.6.2}$$

As in Sect. 2.3 the scatterer density is denoted by $\sigma(\boldsymbol{x}, t)$ and its Fourier transform (the amplitude) by $A(\boldsymbol{q}, t)$. With the definition of $I(\boldsymbol{q}, t)$ in Eq. 3.2.18 $G^{(2)}(\boldsymbol{q}, \Delta t)$ reads

$$\begin{aligned} G^{(2)}(\mathbf{q}, \Delta t) &= \langle \bar{A}(\mathbf{q}, .)A(\mathbf{q}, .)\bar{A}(\mathbf{q}, . + \Delta t)A(\mathbf{q}, . + \Delta t)\rangle \\ &= \left\langle \int d\boldsymbol{x}_1 \dots d\boldsymbol{x}_4 \sigma(\boldsymbol{x}_1, .)\sigma(\boldsymbol{x}_2, .)\sigma(\boldsymbol{x}_3, . + \Delta t) \right. \\ &\quad \left. \sigma(\boldsymbol{x}_4, . + \Delta t)e^{i\boldsymbol{q}(\boldsymbol{x}_1 - \boldsymbol{x}_2 + \boldsymbol{x}_3 - \boldsymbol{x}_4)} \right\rangle \\ &= \int d\boldsymbol{x}_1 \dots d\boldsymbol{x}_4 \langle \sigma(\boldsymbol{x}_1, .)\sigma(\boldsymbol{x}_2, .)\sigma(\boldsymbol{x}_3, . + \Delta t) \\ &\quad \sigma(\boldsymbol{x}_4, . + \Delta t)\rangle e^{i\boldsymbol{q}(\boldsymbol{x}_1 - \boldsymbol{x}_2 + \boldsymbol{x}_3 - \boldsymbol{x}_4)}. \end{aligned} \tag{3.6.3}$$

This four-point correlation can be simplified under the assumption that the correlations decay sufficiently fast, i.e. there exists a length ξ such that for distances $|\Delta \boldsymbol{x}| \gg \xi$ the correlations have decayed to zero for practical purposes. In particular this assumption implies that the correlation functions factorize:

$$\langle \sigma(x_1, t_1)\sigma(x_2, t_2)\rangle = \langle \sigma(x_1, t_1)\rangle\langle \sigma(x_2, t_2)\rangle \quad \text{for } |x_1 - x_2| \gg \xi \qquad (3.6.4)$$

Now I split the domain of integration into four sets:

$$V_{12;34} := \left\{(x_1, x_2, x_3, x_4) \in \left(\mathbb{R}^3\right)^4 \big| |x_1 - x_2| < \xi \wedge |x_3 - x_4| < \xi\right\},$$
$$V_{14;32} := \left\{(x_1, x_2, x_3, x_4) \in \left(\mathbb{R}^3\right)^4 \big| |x_1 - x_4| < \xi \wedge |x_3 - x_2| < \xi\right\},$$
$$V_{13;24} := \left\{(x_1, x_2, x_3, x_4) \in \left(\mathbb{R}^3\right)^4 \big| |x_1 - x_3| < \xi \wedge |x_2 - x_4| < \xi\right\}, \quad (3.6.5)$$

and V' the complement of the union of those.

For points in V' there is obviously an i such that x_i is distant from the other three x_j, so by the four-point analogon of Eq. 3.6.4 the expected value of the four-point product in Eq. 3.6.3 can be split into the product of the expected value of a product of three and the expected value of $\sigma(x_i, .)$, which is a constant, namely the concentration. Performing the Fourier transform with respect to x_i equates to 0 for $q \neq 0$, so V' need not be considered in Eq. 3.6.3.

For the contribution of $V_{13;24}$ in Eq. 3.6.3. I make the substitution $x_3 = x_1 + \Delta x_1$ and $x_4 = x_2 + \Delta x_2$, again using factorization and the definition of G in Eq. 2.3.17 leads to

$$\int dx_1 d\Delta x_1 dx_2 d\Delta x_2 G(\Delta x_1, \Delta t)G(\Delta x_2, \Delta t)e^{iq(2x_1 + \Delta x_1 - 2x_2 - \Delta x_2)}, \quad (3.6.6)$$

which obviously again gives 0 for $q \neq 0$ after integrating over dx_1 or dx_2. Therefore only $V_{12;34}$ and $V_{14;32}$ need to be considered, by doing the appropriate substitutions Eq. 3.6.3 reads

$$G^{(2)}(q, \Delta t) = \int_{V_{12;34}} dx_1 d\Delta x_1 dx_3 d\Delta x_3 G(\Delta x_1, 0)G(\Delta x_3, 0)e^{-iq(\Delta x_1 + \Delta x_3)}$$
$$+ \int_{V_{14;32}} dx_1 d\Delta x_1 dx_3 d\Delta x_3 G(\Delta x_1, \Delta t)G(\Delta x_3, -\Delta t)e^{-iq(\Delta x_1 + \Delta x_3)}.$$

$$(3.6.7)$$

As $G(\Delta x, \Delta t)$ is constant for $|\Delta x| \gg \xi$ and for all Δt, the integrations can be again extended over the whole domain. With the definition of the coherent intermediate scattering function this finally leads to

$$g^{(2)}(q, \Delta t) = \frac{I(q, 0)^2 + I(q, \Delta t)^2}{I(q, 0)^2} = 1 + \left(\frac{I(q, \Delta t)}{I(q, 0)}\right)^2. \qquad (3.6.8)$$

In the literature this is often called the Siegert relation, which is normally written via the normalized amplitude correlation function (or auto-correlation function of first order)

$$g^{(2)}(q, \Delta t) = 1 + \left(g^{(1)}(q, \Delta t)\right)^2. \qquad (3.6.9)$$

There are three points to be noted: First, the dividing of the domain of integration into subdomains and the subsequent factorizations did not take into account that there is a subdomain V_{1234} where all four x_i are close. It is clear, however, that also on V_{1234} the integrand in Eq. 3.6.3 is bounded and that the relative contribution of V_{1234} to the integral in Eq. 3.6.3 becomes negligible when the size of the sample gets much bigger than the correlation length ξ.

Second, in the above derivation the instantaneous intensity as the squared modulus of the amplitude, which itself is given by the Fourier transformation of the scatterer density, is the quantity of interest. However, in the experiment this instantaneous intensity is detected in quanta, and in X-ray physics (contrary to most optical measurements) this effect is not negligible, see Chap. 5. Given an instantaneous intensity $I(q, t)$, the number of photons detected in the time interval dt is a Poisson-distributed random variable with expected value $I(q, t)dt$. Let now $p(I_1, I_2)$ be the joint probability distribution for the instantaneous intensities at a fixed q and at times t_1 and t_2, and let $p(n_1, n_2 | I_1, I_2)$ be the joint probability distribution for the detected number of photons at this times with given intensities I_1 and I_2. The crucial point now is that for $t_1 \neq t_2$ the quantization is uncorrelated, i.e. the probability distribution factorizes:

$$p(n_1, n_2 | I_1, I_2) = p(n_1 | I_1) p(n_2 | I_2) \tag{3.6.10}$$

Therefore correlating the actual numbers of detected photons

$$
\begin{aligned}
\langle n_1 n_2 \rangle &= \int dI_1 dI_2 \, p(I_1, I_2) \int dn_1 dn_2 \, p(n_1, n_2 | I_1, I_2) n_1 n_2 \\
&= \int dI_1 dI_2 \, p(I_1, I_2) \int dn_1 \, p(n_1 | I_1) n_1 \int dn_2 \, p(n_2 | I_2) n_2 \\
&= \int dI_1 dI_2 \, p(I_1, I_2) I_1 I_2 (dt)^2 \\
&= \langle I_1 I_2 \rangle (dt)^2 \tag{3.6.11}
\end{aligned}
$$

is equivalent to correlating the instantaneous intensity, justifying Eq. 3.6.1.

Third, in actual experiments the incoming wave is not an ideal flat, monochromatic wave and the detector has a finite aperture, which can be pictured as if the sample is illuminated by a number of plane waves with no phase relation and therefore with no interference effects on accessible timescales between them. This leads to a partial washing-out of the interference pattern, which is treated in detail in Sect. 6.2. The net effect is just that the magnitude of the term in the measured auto-correlation function due to interference is diminished:

$$g^{(2)}(q, \Delta t) = 1 + \beta \left(\frac{I(q, \Delta t)}{I(q, 0)} \right)^2, \tag{3.6.12}$$

where $\beta < 1$ is the so-called coherence factor.

References

1. J. Als-Nielsen, D. McMorrow, *Elements of Modern X-ray Physics* (Wiley, Chichester, 2001)
2. M. Eisenstadt, A.G. Redfield, Nuclear spin relaxation by translational diffusion in solids. Phys. Rev. **132**, 635 (1963)
3. B. Schönfeld, Local atomic arrangements in binary alloys. Prog. Mater. Sci. **44**, 435 (1999)
4. C.A. Sholl, Diffusion correlation factors and atomic displacements for the vacancy mechanism. J. Phys. C **14**, 2723 (1981)
5. M. Sutton, X-ray intensity fluctuation spectroscopy, in *Neutron and X-ray Spectroscopy* ed. by F. Hippert, E. Geissler, J.L. Hodeau, E. Lelièvre-Berna, J.R. Regnard (Springer, Netherlands, 2006), pp. 297–318

Chapter 4
Characteristics of Diffusion in Selected Systems

This chapter presents the systems which we have either measured in the course of this thesis (see Chap. 7) or are planning to measure. Each system also serves as a prototype to discuss general aspects of solid-state diffusion.

4.1 An Open System: $Si_x Ge_{1-x}$ with the Diamond Lattice

The semiconductors Si and Ge (and also C under special conditions) crystallize in the diamond lattice. This lattice is not a Bravais lattice; it can be constructed from two face-centred cubic lattices translated with respect to each other by [1/4, 1/4, 1/4], therefore there are two crystallographically inequivalent sites in the primitive cell, see Fig. 4.1. Another way to obtain this lattice is to imagine a $2 \times 2 \times 2$ cubic supercell of the body-centred cubic lattice and to take out every other bcc cell. This does not change the nearest neighbour-distance, so the volume fill factor drops by a factor of two compared to the bcc lattice. This large free volume is already an indication why metals do not crystallize in the diamond lattice as a metal tends to pack its atoms as close as possible without significant overlap of the inner shells. Such a reasoning does not hold, however, for the semiconductors with their covalent bonds, rather the diamond lattice leading to to the sp^3-hybridization is the natural choice here.

The large free volume also affects the diffusive behaviour of the atoms: contrary to metals, where it is commonly accepted that vacancies are indispensable for self-diffusion, here also diffusion mechanisms based on self-interstitials are conceivable. The relevant mechanisms are therefore

- the vacancy mechanism: This is just the mechanism which is generally thought to be responsible for self-diffusion in metals—a vacancy hops through the crystal, which leads to nearest-neighbour jumps of the atoms.
- the interstitial mechanism: This means that diffusion happens via jumps of self-interstitials from one interstitial site to the next. Contrary to metals this mechanism is plausible in the diamond lattice as here first there is free volume, therefore a self-interstitial is not prohibitively costly in energy, and second the interstitial sites

M. Leitner, *Studying Atomic Dynamics with Coherent X-rays*,
Springer Theses, DOI: 10.1007/978-3-642-24121-5_4,
© Springer-Verlag Berlin Heidelberg 2012

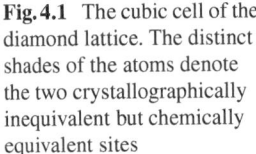

Fig. 4.1 The cubic cell of the diamond lattice. The distinct shades of the atoms denote the two crystallographically inequivalent but chemically equivalent sites

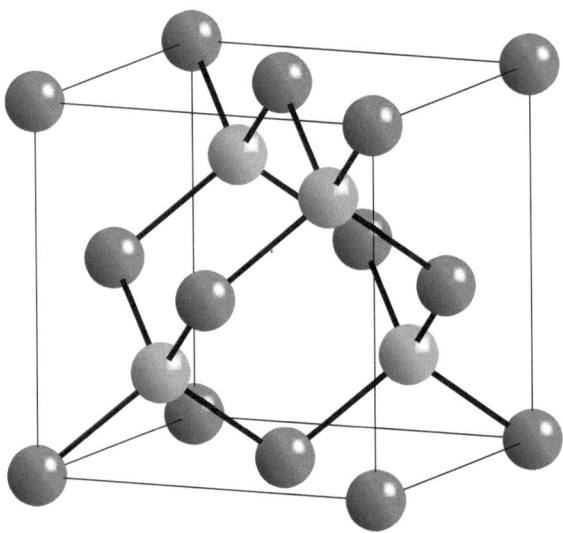

are interconnected with a distance between them equal to the nearest-neighbour-distance of the regular sites.

- the interstitialcy mechanism: This mechanism also relies on self-interstitials, but contrary to the former mechanism here an atom does not jump from interstitial site to interstitial site until it is eventually incorporated again into the crystal, rather an interstitial atom pushes one of its neighbours away and occupies its site, with the pushed-out atom now residing on another interstitial site.

Si and Ge are chemically very similar, therefore the ordering energies in a Si–Ge compound are very small (below 1 meV due to de Gironcoli et al. [7]). If one also assumes the dynamical behaviour of the constituents in a Si–Ge solid solution to be equal—which is not necessarily the case due to the affinity of the vacancies to Ge [22]—the pair-correlation function becomes equivalent to the self-correlation function as the interaction potential in Eq. 2.3.15 vanishes. Therefore a coherent scattering method on a Si–Ge sample measures the same thing as an incoherent method on an elemental sample. Consequently I will confine myself to the elemental case from now on.

An illustration of the vacancy mechanism is given in Fig. 4.2. As the vacancy migrates through the crystal the atoms perform nearest-neighbour jumps from one sublattice to the other (apart from correlation effects as described in Sect. 3.5). Therefore the decay of the intermediate scattering function will be given by two exponentials in the non-degenerate case.

The interstitial mechanism is illustrated in Fig. 4.3. An atom goes to an interstitial site, performs a large number of jumps, and is eventually incorporated into the crystal again. This leads to large effective displacements with an approximately isotropic distribution.

Fig. 4.2 The vacancy diffusion mechanism. The *cube* denotes the vacancy

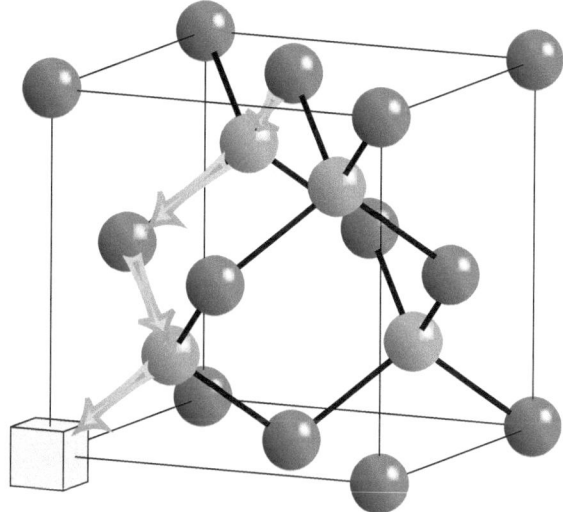

Fig. 4.3 The interstitial diffusion mechanism. The *light grey spheres* are the interstitial sites temporarily occupied by the diffusing atom

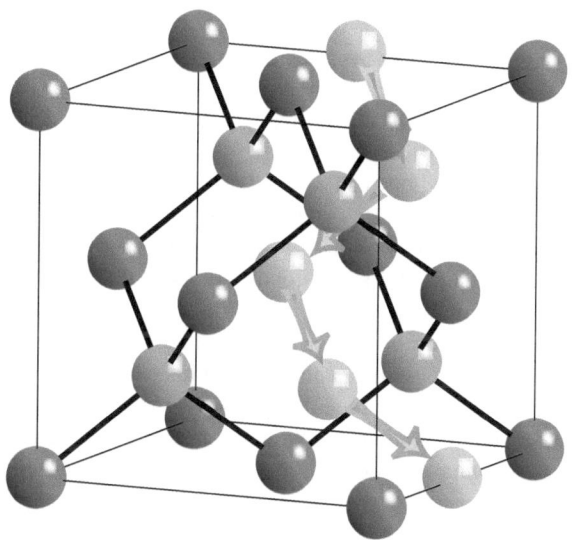

For the interstitialcy mechanism note first that just as the crystal is made up of two sublattices α_1 and α_2, also the interstitial sites can be classified into two sublattices β_1 and β_2. This can be most easily seen by constructing the diamond lattice from the bcc lattice and leaving out every other bcc cell, because then it is obvious that the interstitial sites can be obtained by translating the regular sites by half of a cubic translation vector. Inspecting Fig. 4.4 shows that the nearest-neighbour interstitial sites to an atom sitting on α_2 belong exclusively to β_1 and vice versa. This holds also for α_1 and β_2. Therefore in the interstitialcy mechanism an atom starting out

Fig. 4.4 The interstitialcy
diffusion mechanism. The
light grey spheres are the
interstitial stepping stones
occupied by the diffusing
atoms

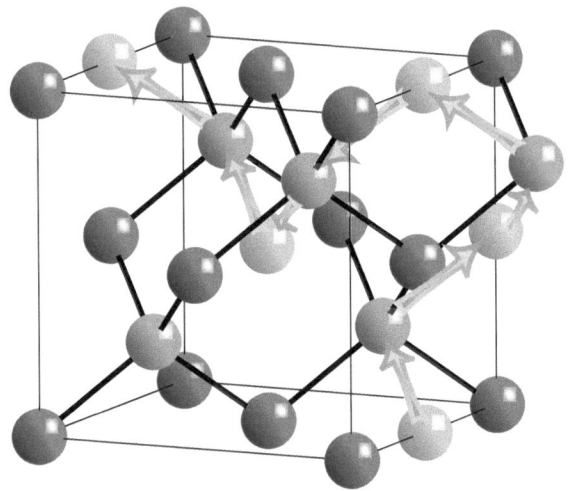

from one sublattice will always end up on this sublattice again, more specifically the
effective exchange vectors are the 12 nearest-neighbour vectors of the face-centred
cubic lattice. So even though the diamond lattice is no Bravais lattice, the intermediate
scattering function decays with a single exponential, because there is no flux between
the two sublattices.

Concluding one can expect to be able to comfortably discern between the various
mechanisms by means of the incoherent (or equivalently coherent) intermediate scattering functions measured for different positions in reciprocal space: the interstitial
mechanism would show no variations in the decay time as a function of the position in reciprocal space apart from a parabolic form at very small scattering angles
(which would allow to determine the average length of the effective translations),
the interstitialcy mechanism would have constant intermediate scattering functions
(i.e. infinite decay times) at the Bragg reflections of the face-centred cubic lattice
which are forbidden in the diamond lattice, and the vacancy mechanism would show
a two-component decay corresponding to nearest-neighbour jumps.

4.2 A Triple Defect System: Ni-Rich B2 NiAl

B2 NiAl is a very well-ordered system. This is because for a stoichiometric composition certain defects are very costly in configuration energy: both a vacancy on
the Al-sublattice and an Al-atom on the Ni-sublattice have an effective formation
energy equal to or more than 2 eV [20]. This excludes thermal Schottky defects or
antisite pairs. The dominant thermal defect therefore consists of three point defects:
a Ni- and an Al-atom leave their position and go to the surface, enlarging the crystal
by one unit cell, and another Ni-atom fills the resulting vacancy on the Al-sublattice,

so effectively two Ni-vacancies and one Ni-antisite are created. This is called a triple defect, costing a bit more than 2 eV [15, 20]. This scenario also holds for the Ni-rich side, where the surplus Ni-atoms are accommodated as antisites on the Al-sublattice. On the Al-rich side the situation looks different, however. As an Al-antisite is very costly, the ground state has structural vacancies on the Ni-sublattice instead. Here the dominating thermal defect is the creation of an Al-antisite with the concomitant annihilation of two vacancies on the Ni-sublattice, an inverse triple defect of some sort, leading to the counterintuitive feature of decreasing vacancy concentration with increasing temperature [20].

This high degree of order with a thermal defect concentration less than 10^{-3} at temperatures as high as 1300 K [15] imposes severe restrictions on the dynamics in the system. The vacancy (or more general, the diffusion vehicle) has to choose its path through the crystal so that the necessary disturbances of the order are only temporary, it has to restore the order again upon leaving. In the following I will give examples of such mechanisms for the Ni-rich case and show how these mechanisms manifest themselves in the coherent and incoherent intermediate scattering functions obtained by Monte Carlo simulations.

The fact that the vacancies are preferably located on the Ni-sublattice allows for some preliminary observations to be made: as soon as enough Ni-antisites are available the tracer diffusivity of Ni and Al will decouple as this enables the vacancies to move by nearest-neighbour jumps via the Ni-antisites without disturbing the order. Even without a significant number of Ni-antisites diffusion of Ni could happen via next-to-nearest-neighbour jumps, i.e. jumps along an edge of the primitive cubic cell. The activation energy for such a jump is with a value of about 2.5 eV quite high, but as it induces no disorder and only needs one vacancy it cannot be excluded a priori [21].

Both of the above processes only lead to Ni diffusion, however. Concerning Al diffusion two mechanisms have been proposed:

- six-jump cycles [10]: a vacancy on the Ni-sublattice performs one of a number of sequences of six nearest-neighbour jumps, where the latter three undo the disordering caused by the former three, leading to a displacement of the vacancy, one Ni-atom, and two Al-atoms.
- the triple-defect mechanism [26]: this mechanism involves a localized triple defect, that is two Ni-vacancies and one Ni-antisite next to each other. Concerted nearest-neighbour jumps of the two vacancies lead to the displacement of two Ni-atoms, one Al-atom, and the defect, incurring only a low amount of additional disorder during the transition.

In a six-jump cycle the vacancy moves along a path $A \rightarrow B \rightarrow C \rightarrow D \rightarrow A \rightarrow B \rightarrow C$ connecting nearest neighbours, where A and C are on the Ni-sublattice and B and D on the Al-sublattice. After the six jumps have been executed the vacancy has moved to the position C, the Ni-atom initially on C is now on A, and the Al-atoms on B and D have exchanged place. After accounting for cubic symmetry there are three possible choices of the path: in the so-called $\langle 110 \rangle$ cycle C is at a position $\langle 110 \rangle$ relative to A and B to D is therefore a vector $\langle 001 \rangle$ (this variant is illustrated

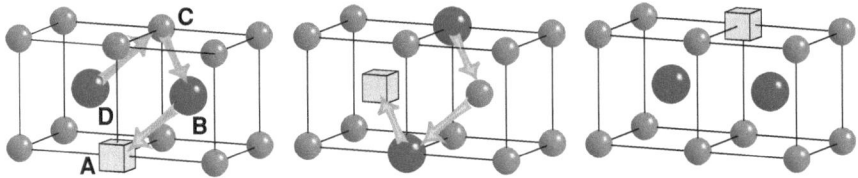

Fig. 4.5 The $\langle 110 \rangle$ variant of the six-jump cycle. The small atoms are Ni, the large Al, and the vacancy is the *cube*. On the *left* is the initial state, in the *middle* the intermediate configuration corresponding to the maximum in energy along the transition path, on the *right* the final state

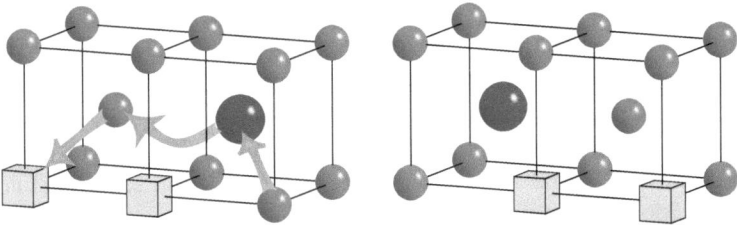

Fig. 4.6 A triple-defect jump as it is commonly pictured with the initial state on the *left*, final state on the *right*. Note that the actual initial and final position of the moving vacancy is inconsequential

in Fig. 4.5), the straight $\langle 100 \rangle$ cycle has A and C displaced by $\langle 100 \rangle$ and B and D by $\langle 011 \rangle$, and the bent $\langle 100 \rangle$ cycle again has A and C displaced by $\langle 100 \rangle$, but B and D by $\langle 010 \rangle$. Straight and bent refers to whether the path lies in a plane or not. It is commonly accepted that the $\langle 110 \rangle$ cycle is the energetically most favoured variant with a saddle point energy slightly smaller than 3 eV [17, 21].

It has been stated above that the triple defect is commonly thought to be the dominant thermal defect in stoichiometric and Ni-rich NiAl. The triple defect mechanism as illustrated in Fig. 4.6, however, needs the three point defects to be located on neighbouring sites. At low defect concentrations (as it is the case in NiAl) this obviously lowers the entropy by a large amount compared to the free case, or equivalently the instances where the three defects are in a position so as to initiate a move are very rare. Xu and Van der Ven [28] give quantitative results where they show that there are favoured arrangements of the triple defect leading to a higher concentration compared to the mean-field result (i.e. neglecting interactions), although they still stay rare. Once there is a localized triple defect, however, its migration energy is only slightly larger than 1 eV.

The question of chemical diffusion necessitates further considerations: In any real sample the dominant defects will be of structural nature due to the inevitable off-stoichiometry as opposed to thermal defects. In the Ni-rich case this means that the number of structural Ni-antisites will be much higher than the number of thermal vacancies, let alone localized triple defects. The elementary event of chemical diffusion is therefore the exchange of the occupation of two Al-sites: one Ni-antisite becomes Al, and one Al becomes a Ni-antisite. Note that it is advisable to think in

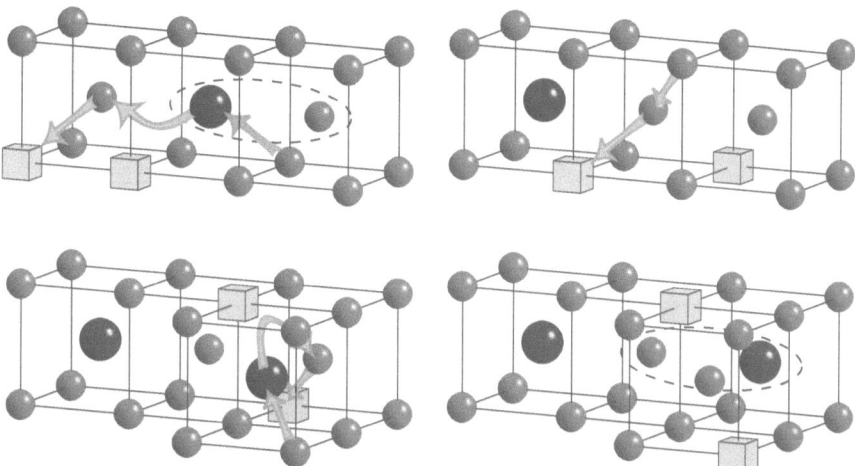

Fig. 4.7 The elementary event of chemical diffusion in the case of diffusion by the triple-defect mechanism. A triple defect enters the neighbourhood of a Ni-antisite, the vacancies rearrange, the triple defect formed with the encountered Ni-antisite leaves, resulting in the apparent translation of the antisite to a neighbouring Al-site (*highlighted*)

a sufficiently coarse temporal resolution: picture a localized triple defect diffusing through the crystal with a very small off-stoichiometry. After each step the triple defect's Ni-antisite occupies a different Al-site, but these are not events of chemical diffusion, because even if one triple defect makes an arbitrary number of steps, the overwhelmingly larger number of structural Ni-antisites is not affected. For chemical diffusion to happen, that is for the arrangement of structural Ni-antisites to change, it is necessary that the triple defect enters a neighbouring cell of a Ni-antisite, the vacancies change over to the encountered Ni-antisite forming again a localized triple defect, and that this new triple defect diffuses away, leaving the Ni-atom the former triple defect brought with it stranded on an Al-site (see Fig. 4.7). Such a change-over of the vacancies corresponds to the elementary event of chemical diffusion. An analysis of the possible apparent translation vectors will be given below.

The existence of structural Ni-antisites in a concentration greater than the concentration of thermal defects also has an effect on diffusion via the six-jump cycle which apparently has not yet been treated in the literature: it can lead to what I will call the 4 + 2-jump mechanism (pictured in Fig. 4.8). Start with a vacancy which has exactly one Ni-antisite among its eight nearest neighbours. Let this vacancy initiate a six-jump cycle where the Ni-antisite is on the position D of the path in the above notation. After the second jump of the vacancy, however, with an Al- and a Ni-antisite created, the next two jumps of the vacancy again restore the order. The vacancy can now reach any of the seven other Ni-neighbours of the Ni-antisite via two nearest-neighbour jumps without producing disorder. Again it has a Ni-antisite as a nearest neighbour and can therefore start the next four-jump cycle. Just as in the case of the triple-defect mechanism a more complicated defect (involving more point defects) enables

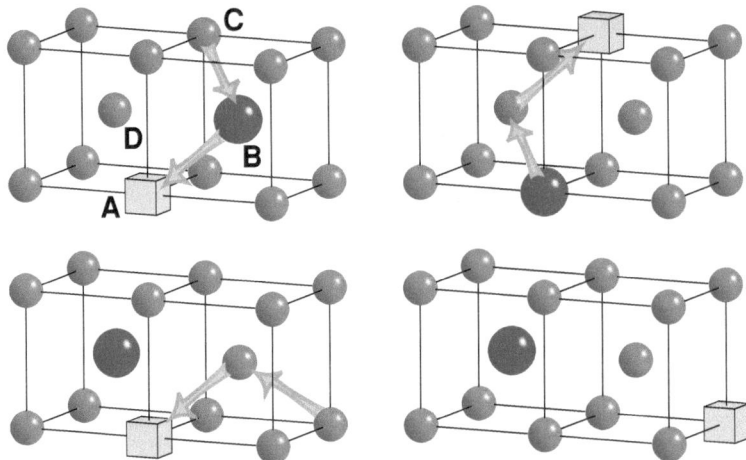

Fig. 4.8 The 4+2-jump mechanism. The jumps leading to the movement of the defect are in the *upper row*, the *lower left* shows the setting-up for the next jump

a diffusion path with a smaller migration energy, but here the second point defect is not thermal but structural and therefore available in much greater concentrations: already in $Ni_{51}Al_{49}$ 15% of the vacancies have at least one Ni-antisite among their nearest neighbours in the mean-field approximation. Calculations of the migration energies of the 4+2-jumps would be desirable. Chemical diffusion effected by the 4+2-jump mechanism works analogously as in the case of the triple-defect mechanism: a vacancy and a Ni-antisite arrive in the neighbourhood of another Ni-antisite, the vacancy changes over and leaves with the second Ni-antisite.

For the cases of both the triple-defect mechanism and the 4+2-jump mechanism the elementary event of chemical diffusion is given by a defect's vacancies deserting the Ni-antisite they came with and forming a new defect with another Ni-antisite. When this new defect leaves the vicinity, the Al-site formerly occupied by Ni becomes occupied by Al. The Al-antisite therefore apparently jumps in the reverse direction the vacancies took. The flow of the vacancies can already happen when the Ni-antisites' coordination cubes share only a corner. It is therefore plausible that the apparent Ni-antisite jumps are along $\langle 100 \rangle$, $\langle 110 \rangle$, and $\langle 111 \rangle$. Correlation effects as in the encounter model (see Sect. 3.5) will yield contributions from farther jumps. This makes an analytic calculation of the probabilities of the respective jumps infeasible, I will instead present results from simulations below.

The case presented above was for a small off-stoichiometry: defects diffuse far before they encounter a Ni-antisite, the vacancies exchange, and the new defect diffuses away. Once the off-stoichiometry gets larger, however, the Ni-antisites become interconnected. The relevant percolation threshold is at around 10% of Al-sites occupied by Ni [9] (R. Ziff private communication), that is a Ni-concentration of 55% neglecting thermal defects. If the diffusing defect now encounters a cluster of Ni-antisites, it can exit far from where it entered. In the extreme case where

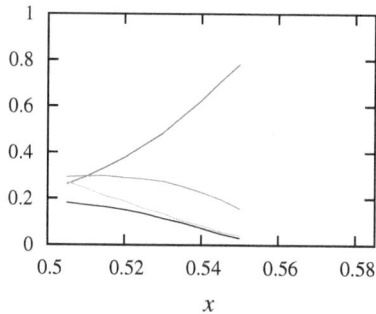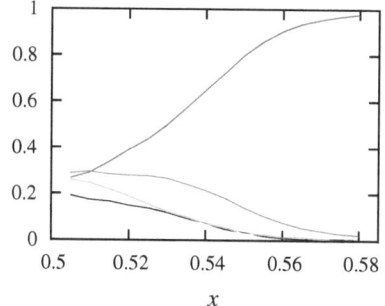

Fig. 4.9 Weights of the apparent exchange vectors of chemical diffusion for the triple-defect mechanism (*left*) and the 4+2-jump mechanism (*right*) as a function of Ni-concentration x: $\langle 100 \rangle$ is *red*, $\langle 110 \rangle$ *green*, $\langle 111 \rangle$ *blue*, and farther jumps *pink*

practically each Ni-antisite belongs to an infinite cluster the analysis of the apparent displacements becomes easy: the vacancy diffuses through the cluster (or the vacancies diffuse independently through the cluster and meet somewhere in the case of the triple-defect mechanism), it makes one 4-jump cycle (or a triple-defect jump) and it diffuses away via the Ni-antisites. As a consequence in this case the apparent displacements are equal to the actual displacements of the Ni-atoms, i.e. $\langle 100 \rangle$-jumps as is commonly accepted.

Monte Carlo simulations show that above considerations are qualitatively correct: the relative weights of exchanges into the first seven shells of Al-sites are given in Fig. 4.9 as a function of Ni-concentration x for both processes, where the shells four to seven are combined for simplicity. Further jumps can be neglected. The simulations used a box of $32 \times 32 \times 32$ B2-cells. They were run in the limit of low temperature (as is appropriate for NiAl): there were no thermal defects apart from either one or two vacancies (depending on the mechanism). Energetical interactions between the defects were not considered, i.e. no short-range order was present. The vacancies could either make nearest-neighbour jumps onto sites occupied by Ni or perform the correlated jump sequence (if allowed by the configuration of the vacancies in the case of the triple-defect mechanism). For the triple-defect mechanism the attempt frequency for a valid nearest-neighbour jump sequence via a Ni-antisite was more probable than for a valid correlated jump by a factor of 100, for the 4+2-jump mechanism by a factor of 1,000. This does not mean that there were 100 antisite jumps per triple-defect jump, rather there were much more, as for higher x the vacancies were only very rarely in a configuration allowing a correlated jump, being dispersed over a cluster. This was the reason why the triple-defect mechanism was simulated only up to a Ni-concentration of $x = 55\%$, higher x would need much more CPU time. For the 4+2-jump mechanism, however, the ratio of performed jumps is in first order equal to the ratio of attempt frequencies.

The results in Fig. 4.9 show that the weights of the apparent exchanges into the first three shells are of the same order of magnitude for small off-stoichiometry, but

Fig. 4.10 Tracer diffusivities of Ni (*red*) and Al (*blue*) for both the 4 + 2-jump mechanism (*solid lines*) and the triple-defect mechanism (*dashed lines*) normalized by the chemical diffusion constant

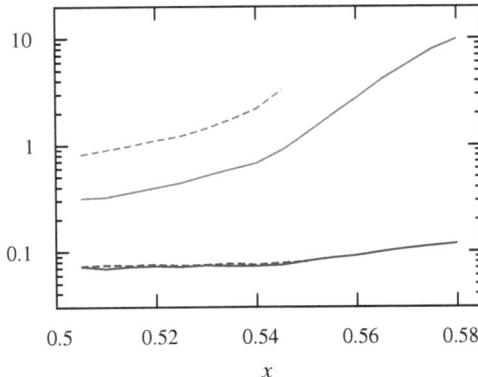

exchanges into further shells also contribute. For Ni-concentrations x larger than 55%, where more and more Ni-antisites belong to an infinite cluster, the exchange along $\langle 100 \rangle$ dominates. They also show that the two mechanisms do not significantly differ in the weights of the apparent displacements. This becomes easily understandable if one considers the defect as an abstract entity irrespective of its actual configuration. Both the triple defect and the bound vacancy-antisite pair of the 4 + 2-jump mechanism contain one Ni-antisite. Let the position of the abstract defect on the lattice be given by the B2-cell in which the Ni-antisite is located. The defect can move along $\langle 100 \rangle$-directions by a correlated jump sequence, displacing one Al-atom in the opposite direction. If one of the 26 Al-sites next to the defect is occupied by a structural Ni-antisite, the defect can migrate to this position by nearest-neighbour jumps of Ni. If the probability for this migration over Ni-antisites is much higher than for the jump sequences (as will be the case due to both energy and entropy) each position in the cluster of Ni-antisites will be equally probable as a starting point for the next correlated jump sequence. In this abstract description the actual configuration of the defect was not mentioned, so they are equivalent from the point of chemical diffusion. Needless to say, energetical interactions between the point defects would disturb the equivalence.

Even though the two mechanisms do not differ in the chemical diffusion they promote (i.e. the pair-correlation of the system), the self-correlation is distinct. Figure 4.10 shows the tracer diffusivities normalized by the chemical diffusion constant for both constituents and both mechanisms. The diffusivity of Al is very similar for both processes, which can be understood by an argumentation as in the last paragraph. Ni, on the other hand, diffuses much faster under the triple-defect mechanism than under the 4 + 2-jump mechanism. This is mostly due to the fact that the two vacancies will shuffle the seven Ni-atoms touching the cube of the triple defect very effectively by a succession of two nearest-neighbour jumps as pictured in the right top panel of Fig. 4.7. Once the percolation threshold at $x = 0.55$ is reached the diffusivity of Ni is bounded only by the size of the simulation box due to the calculation of the self-correlation function as implemented here.

Concluding it seems that it is impossible to decide between diffusion mechanisms by a coherent scattering experiment alone. Combining such measurements with a determination of the tracer diffusivities or comparing the vacancy concentration as a function of composition with the measured chemical diffusivities should allow to decide, however.

4.3 A Short-Range Ordered System: $Cu_{90}Au_{10}$

Cu–Au is a classical system in metal physics. For one it exhibits the prototypical examples of the $L1_0$- and the $L1_2$-phases with CuAu and Cu_3Au, respectively, it has provided the inspiration for the Bragg–Williams model of long-range order [2] and its refinement by Shockley [25] to include nearest-neighbour interactions, thereby laying the groundwork for the treatment of disorder in alloys as it is nowadays done by cluster expansion, and also the quantitative description of short-range order by Cowley [6] was introduced for measurements on this system, to name just a few fundamental contributions to physics.

The aspect I want to treat here is the influence of the short-range order on the atomistic dynamics. Due to Massalski [18] the solubility of Au in Cu exceeds 10 at.% for all temperatures where equilibration is experimentally feasible, but the proximity of the $L1_2$-phase of Cu_3Au at low temperatures suggests the emergence of short-range order. This is proven by the experimental investigations of Schönfeld et al. [24]: in $Cu_{90}Au_{10}$ at 573 K short-range order, but no long-range order is found. Specifically the probability for a given nearest-neighbour site of a Au atom to be occupied by Au is only 2% compared to the mean-field value of 10%, the probability for a $\langle 100 \rangle$-Au neighbour is 17% on the other hand. Correlations over longer ranges are already very weak. As expected, these deviations from the mean-field value correspond qualitatively to the occupations in the $L1_2$-phase.

As the Cu–Au alloy is a close-packed metal, the diffusion mechanism will be ordinary nearest-neighbour jumps into vacancies. The influence of short-range order on the coherent intermediate scattering function was treated in Sect. 2.3 in the high-temperature limit. In order to explore the range over which the high-temperature limit is applicable and to qualitatively investigate what happens outside this range I conducted simulations.

The stochastic simulation of the trajectory of a system through configuration space is commonly known as Monte Carlo simulation. The necessary ingredients for such a simulation are an initial configuration and a rule specifying the transition rates for the system to change from one configuration to the next. For simulations on a discrete timescale these rates can equivalently be given as the transition probabilities per unit time. When simulating static equilibrium properties the only requirement on the transition rates is that they fulfil detailed balance, in fact this is also a sufficient requirement for the system to converge to thermodynamic equilibrium provided the configuration space does not decompose into unconnected domains. Among the abundance of rules the Metropolis rule [19] is the most popular. For the simulation of

Fig. 4.11 Sketch of the energy landscape corresponding to the Metropolis rule (*red*) and the midpoint rule (*blue*)

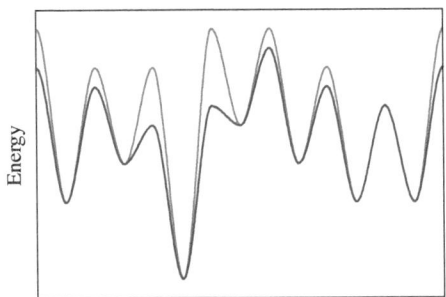

dynamics, however, the set of possible transitions and their respective probabilities should mirror the physical model. When simulating vacancy diffusion, for example, one would only allow nearest-neighbour jumps of the vacancies. The transition probabilities are in most investigations given by the energetical difference of initial and final state via the Metropolis or Glauber [12] rule. From classical transition state theory it follows, however, that the energy difference between the initial state and the saddle point of the transition path in the energy landscape governs the transition probabilities via the Boltzmann factor (as in Sect. 2.3), with an additional entropy contribution from the fact whether the saddle point in the energy landscape is wide or narrow [27].

The Monte Carlo simulations to be reported in the following were done with two choices for the transition probabilities: Denote the energies of the initial state and the final state with E_i and E_f, respectively. The probability for transition from the initial to the final state within the unit time interval is then proportional to $e^{-\Delta E/k_B T}$, where ΔE is the difference of the energy of the saddle point configuration and the energy of the initial configuration E_i, given by

- the Metropolis rule: $\Delta E = E_0 + \max(E_f - E_i, 0)$ or
- the midpoint rule: $\Delta E = E_0 + (E_f - E_i)/2$.

An illustration of the energy landscapes along the path of the vacancy corresponding to these rules for given energies of the stable positions is shown in Fig. 4.11. Note that a rule yielding the Glauber probabilities from the Boltzmann factor of a saddle point energy cannot be devised as this energy would have to be temperature-dependent, in fact the Glauber probabilities converge to the Metropolis probabilities for low temperatures and to the midpoint probabilities for high temperatures.

The simulations were done for an Ising model inspired by the short-range ordered solid solution $Cu_{90}Au_{10}$ as investigated by Schönfeld et al. [24]: the system was a face-centred cubic lattice occupied by 90% A-particles and 10% B-particles and one vacancy. Interactions between the particles were considered for the first two shells, with the respective effective interaction energies J and $-0.256J$ (this corresponds to the ratio between the experimental values of V_{110} and V_{200}), the vacancy was not considered to interact.

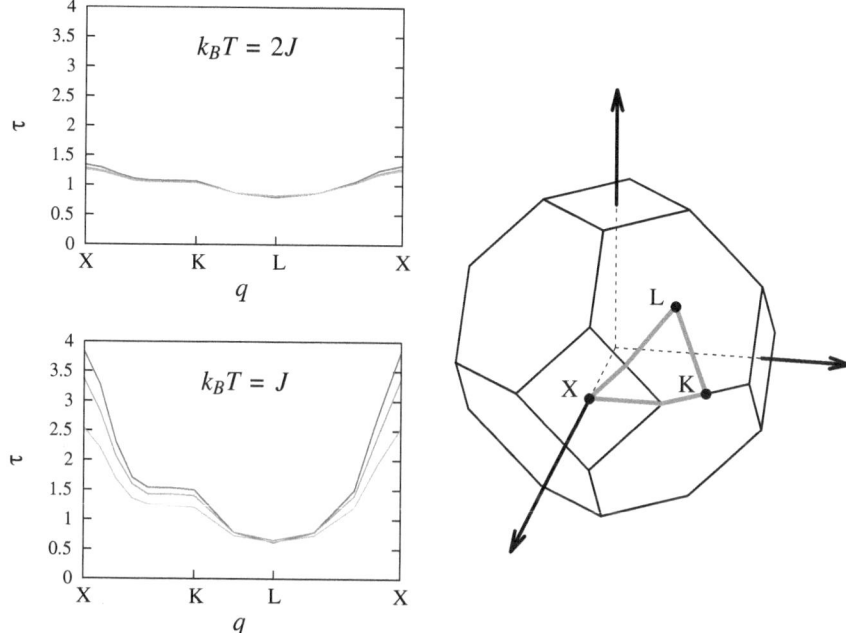

Fig. 4.12 Correlation time τ as a function of $\boldsymbol{q}$ for the Metropolis rule (*red*), midpoint rule (*blue*), and theoretical expression (*green*). *Upper panel* for medium temperature, *lower panel* for low temperature. Path through reciprocal space on the Brillouin zone of the face-centred cubic lattice on the *right*

The resulting fitted decay times of the coherent intermediate scattering function are given in Fig. 4.12. For a temperature of $k_B T = 2J$ the decays could be well described by exponential decays and the approximations of Sect. 2.3 seem to hold: the simulations with the two transition rules agree very well with each other and with the value predicted by Eq. 2.3.21. At a temperature of $k_B T = J$ (with the energies of Cu$_{90}$Au$_{10}$ this corresponds to 530 K) short-range order is much stronger (visible in the more pronounced features of $\tau(\boldsymbol{q})$), the decay shows deviations from a single exponential (not shown), and when fitting with a single exponential the fitted decay times display systematic deviations both from the theoretical value and between the two models. This demonstrates that the pair-correlation function (and equivalently the coherent intermediate scattering function) of a short-range ordered system is sensitive to the relationship between the local atomic configuration and the transition frequencies. Here the saddle-point energies (and therefore the transition frequencies) were modelled only as a function of the energy of the initial and the final configuration, but there are endeavours to calculate these energies from first principles for arbitrary local neighbourhoods of the jumping atoms (e.g. Ramanarayanan et al. [22]). As much as so-called ab-initio computations are accepted today, in the majority of these investigations the energy of stable (or even ground-state) configurations is the quantity of interest. Such quantities can also be verified conveniently in the experiment. Configurations along the jump path, however, cannot be examined by static experimental methods, as a given atom is only a negligible amount of time involved in a jump. Measurements of the coherent intermediate scattering function therefore seem to be the only possibility for experimentally verifying such calculations. This is not an irrelevant point as algorithms (or pseudopotentials) accepted to work for stable configurations a priori need not work for configurations far from the energetical minimum.

4.4 A Model of an Amorphous System

Up to here this thesis dealt with diffusion in crystalline systems. Now I want to go beyond these limits and consider diffusion in amorphous media. This is a really bustling field, especially in the XPCS-community, where the vast majority of publications report experiments on soft condensed matter (mostly gels or colloidal glasses). This is probably due to the fact that these systems scatter strongly in the small-angle regime, allowing such experiments to be performed without much effort. It has been proposed (and it is now widely accepted) that the dynamics observed in all these systems conform to a universal principle [5], which has been termed "jamming" [16].

The other main direction of approach to this problem is via computer simulations. There is a number of works which simulate a glass-forming system with various choices of the underlying dynamics [1, 13]. Mostly these are model systems with Lennard-Jones-potentials or, even simpler, hard-sphere potentials. The results from these simulations is by default presented in terms of the incoherent intermediate scattering function. This is motivated by the claim that in a colloidal glass with a polydispersity in the refractive indices of the particles it is in principle possible to study the incoherent intermediate scattering function by (X)PCS. This can be achieved by choosing the refractive index of the medium equal to the mean refractive index of the particles, with an argumentation as in Sect. 3.4 the coherent scattering cross section of the particles is then equal to zero. I think another reason for only considering the self-correlations is that it is much harder to obtain good statistics for the coherent intermediate scattering function than for the incoherent intermediate scattering function. For a metallic glass, which I will treat in Sect. 7.4, there is no medium allowing to tune its refractive index, which necessitates a description via the coherent intermediate scattering function.

In order to obtain the coherent intermediate scattering function I conducted simulations on a model system. I used the binary mixture of hard spheres proposed by Brambilla et al. [3] with 1,024 particles and periodic boundary conditions. In order to prevent crystallization the particles have different sizes, one half has a radius a factor of 1.4 larger than the other half. The system evolves by Monte Carlo algorithm invented by Metropolis et al. [19] and revived by Berthier and Kob [1], but contrary to these works I use isotropic dynamics by drawing the particle displacements from a sphere as opposed to a cube. The invariance of the position of the centre of mass

Fig. 4.13 Coherent
intermediate scattering
function in the structure peak
(*solid red*). For reference
also compressed exponential
decays $\exp(-(\Delta t/\tau)^\gamma)$ are
given, with $\gamma = 1$ (*dashed
blue*) and $\gamma = 2$ (*dotted
green*)

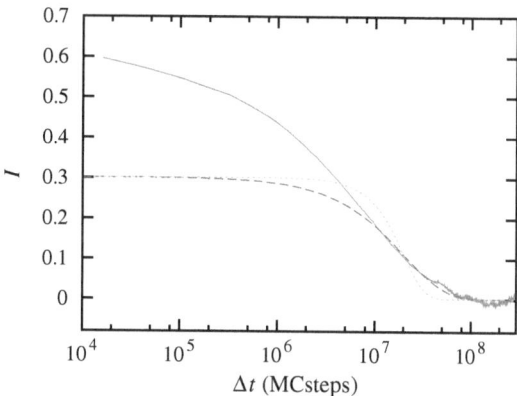

is enforced. The resulting coherent intermediate scattering function (assigning the
same scattering cross section to both kinds of particles) evaluated in the structure
peak with a volume fill factor of 0.59 is given in Fig. 4.13. The timescale is given
in Monte Carlo steps, where one MCstep means 1,024 attempted particle moves
(on average each particle has the possibility to move once).

From this figure it seems that the late stages of relaxation can be well fitted by an
exponential decay; if one wanted to fit the relaxations on a wider range of timescales
one would have to use the form of a stretched exponential. This is also plausible,
as the superposition of independent relaxations on different timescales (each given
by a proper exponential) leads naturally to a stretched exponential. The quantitative
evaluation of Kob and Andersen [14] gives stretching parameters γ on the order
of 0.8. No indication for compressed exponential decays can be inferred from the
simulations presented here or from most other works, only recently such decays seem
to have been observed when doing simulations with many-body potentials [8, 23].

Resuming the review of directions of research into amorphous media, it has to
be stated that besides the great amount of experimental work and computer simulations the theoretical treatments seem to be lacking. I am not aware of a fundamental explanation of the fact that upon entering the jamming regime the correlation
functions start to display a form which is commonly fitted with compressed exponentials [11]. The fact that such compressed exponentials are commonly encountered in experiments on diverse systems and never in simulations can give an
indication, however. There are two main differences between experiments and simulations: The first is the question of accessible timescales. In a metallic glass a
particle explores its cage formed by the neighbouring particles within on the order
of 10^{-12} s (the inverse of the Debye frequency), whereas it leaves this cage after
about 10^3 s. In Fig. 4.13 it can be seen that the particles move on atomic distances
with a timescale of about 10^7 MCsteps, whereas the initial decay (not shown)
happens at about 10 MCsteps. The simulations presented here are in this aspect the
state of the art (the accessible timescales), see for example Berthier and Kob [1].
The other difference is the spatial scale: While simulations are restricted to a

few thousand particles, in reality systems are much larger. It is standard proce-
dure in publications to claim that finite size effects are found to be negligible by
comparing the results from simulations using systems of differing sizes. It has been
proposed, however, that the universal dynamics in jamming systems are due to
relaxations on all lengthscales, leading to so-called heterogeneous dynamics [4].
This is plausible, since independent localized relaxations without memory lead
immediately to an exponential or even stretched exponential decay in time, but can
never cause compressed exponentials. It is obviously not possible to simulate such
processes in the foreseeable future with the present models, where each particle is
treated explicitly.

References

1. L. Berthier, W. Kob, The Monte Carlo dynamics of a binary Lennard-Jones glass-forming
   mixture. J. Phys. Condens. Matter **19**, 205130 (2007)
2. W.L. Bragg, E.J. Williams, The effect of thermal agitation on atomic arrangement in alloys.
   Proc. R. Soc. Lond. Ser. A **145**, 699 (1934)
3. G. Brambilla, D. El Masri, M. Pierno, L. Berthier, L. Cipelletti, G. Petekidis, A.B. Schofield,
   Probing the equilibrium dynamics of colloidal hard spheres above the mode-coupling glass
   transition. Phys. Rev. Lett. **102**, 85703 (2009)
4. L. Cipelletti, H. Bissig, V. Trappe, P. Ballesta, S. Mazoyer, TRC: a new tool for studying
   temporally heterogeneous dynamics. J. Phys. Condens. Matter **15**, S257 (2003)
5. L. Cipelletti, L. Ramos, S. Manley, E. Pitard, D. Weitz, E. Pashkovski, M. Johansson, Universal
   non-diffusive slow dynamics in aging soft matter. Faraday Discuss. **123**, 237 (2003)
6. J.M. Cowley, X-Ray measurement of order in single crystals of CuAu. J. Appl. Phys. **21**,
   24 (1950)
7. S. de Gironcoli, P. Giannozzi, S. Baroni, Structure and thermodynamics of $Si_x Ge_{1-x}$ alloys
   from ab initio Monte Carlo simulations. Phys. Rev. Lett. **66**, 2116 (1991)
8. E. Del Gado, W. Kob, Length-scale-dependent relaxation in colloidal gels. Phys. Rev. Lett. **98**,
   28303 (2007)
9. S.V. Divinski, L.N. Larikov, Diffusion by anti-structure defects in non-stoichiometric intermetallic compounds with B2 and structures. J. Phys. Condens. Matter **9**, 7873 (1997)
10. E.W. Elcock, C.W. McCombie, Vacancy diffusion in binary ordered alloys. Phys. Rev. **109**,
    605 (1958)
11. P. Falus, M.A. Borthwick, S. Narayanan, A.R. Sandy, S.G.J. Mochrie, Crossover from stretched
    to compressed exponential relaxations in a polymer-based sponge phase. Phys. Rev. Lett. **97**,
    66102 (2006)
12. R. Glauber, Time-dependent statistics of the Ising model. J. Math. Phys. **4**, 294 (1963)
13. W. Kob, H.C. Andersen, Testing mode-coupling theory for a supercooled binary Lennard-Jones
    mixture I: the van Hove correlation function. Phys. Rev. E **51**, 4626 (1995)
14. W. Kob, H.C. Andersen, Testing mode-coupling theory for a supercooled binary Lennard-Jones
    mixture II: intermediate scattering function and dynamic susceptibility. Phys. Rev. E **52**, 4134
    (1995)
15. P.A. Korzhavyi, A.V. Ruban, A.Y. Lozovoi, Y.K. Vekilov, I.A. Abrikosov, B. Johansson,
    Constitutional and thermal point defects in B2 NiAl. Phys. Rev. B **61**, 6003 (2000)
16. A.J. Liu, S.R. Nagel, Jamming is not just cool any more. Nature (London) **396**, 21 (1998)
17. K.A. Marino, E.A. Carter, First-principles characterization of Ni diffusion kinetics in β-NiAl.
    Phys. Rev. B **78**, 184105 (2008)

18. T. Massalski (ed.), *Binary Alloy Phase Diagrams* (American Society for Metals, Materials Park Ohio, 1986)

19. N. Metropolis, A.W. Rosenbluth, M.N. Rosenbluth, A.H. Teller, E. Teller, Equation of state calculations by fast computing machines. J. Chem. Phys. **21**, 1087 (1953)

20. B. Meyer, M. Fähnle, Atomic defects in the ordered compound B2-NiAl: a combination of ab initio electron theory and statistical mechanics. Phys. Rev. B **59**, 6072 (1999)

21. Y. Mishin, A.Y. Lozovoi, A. Alavi, Evaluation of diffusion mechanisms in NiAl by embedded-atom and first-principles calculations. Phys. Rev. B **67**, 14201 (2003)

22. P. Ramanarayanan, K. Cho, B.M. Clemens, Effect of composition on vacancy mediated diffusion in random binary alloys: first principles study of the Si_{1-x} Ge_x system. J. Appl. Phys. **94**, 174 (2003)

23. S. Saw, N.L. Ellegaard, W. Kob, S. Sastry, Structural relaxation of a gel modeled by three body interactions. Phys. Rev. Lett. **103**, 248305 (2009)

24. B. Schönfeld, M.J. Portmann, S.Y. Yu, G. Kostorz, The type of order in Cu-10 at.% Au—evidence from the diffuse scattering of X-rays. Acta Mater. **47**, 1413 (1999)

25. W. Shockley, Theory of order for the copper gold alloy system. J. Chem. Phys. **6**, 130 (1938)

26. N.A. Stolwijk, M. van Gend, H. Bakker, Self-diffusion in the intermetallic compound CoGa. Philos. Mag. A **42**, 783 (1980)

27. G. H. Vineyard, Frequency factors and isotope effects in solid rate processes. J. Phys. Chem. Solids **3**, 121(1957)

28. Q. Xu, A. Van der Ven, First-principles investigation of migration barriers and point defect complexes in B2–NiAl. Intermetallics **17**, 319 (2009)

Chapter 5
Data Evaluation

This chapter describes the steps taken in evaluating the data obtained in an XPCS measurement. It aims to be both methodologically thorough and helpful to the reader in reproducing the work presented here, it will therefore be very detailed.

The principal problem is the following: the scattered X-rays are detected with a charge-coupled device (CCD), where one detected photon generates a droplet of charge in a small number of adjacent pixels, for this process see for instance Miyata et al. [2]. Additional charge is generated over time due to leakage current. This chapter describes how to first detect as exactly as possible where each photon hit the chip, and second how to compute the auto-correlation functions from these data.

Data evaluation is done via a suite of command-line programs written in C. The parameters are specified via a common parameter file. The programs are: `evaluatedark` for computing the dark current, `hist` for computing histograms of the ADUs in a droplet, `evaluatedroplets` for converting droplets of generated charge to single photon events, and `computeacf` and `computettacf` for computing the ordinary auto-correlation function and the two-time auto-correlation function, respectively. The workings of these programs are described below. They are available on request.

5.1 Raw Data Files

In an XPCS experiment one measures the number of X-ray photons scattered into a given direction as a function of time. The time resolution needed is dictated by the dynamics in the sample. For optimal statistical accuracy one would in principle want the data of as many equivalent directions as possible, corresponding to many detectors. This leads to a trade-off between many equivalent directions and good temporal resolution: A fast read-out implies a low number of pixels both because of data transfer limitations and because of the fact that fast detectors (such as avalanche photodiodes) are much more complicated and therefore bigger than pixels on a CCD, which are rather slow. For this thesis scattering in the diffuse regime was

M. Leitner, *Studying Atomic Dynamics with Coherent X-rays*,
Springer Theses, DOI: 10.1007/978-3-642-24121-5_5,
© Springer-Verlag Berlin Heidelberg 2012

investigated, entailing the need for a large number of pixels, which can only be obtained with a CCD.

The raw data for a given measurement run consist of a sequence of frames, which means that the electric charge built up in the distinct pixels on the camera's chip is read out every few seconds and stored in a sequence of files. With an exposure time on the order of 10 s and a read-out time of about 1 s a run corresponds to about 500–1,000 frames. Contemporary CCD-cameras have on the order of $1,000 \times 1,000$ pixels. Multiplying these numbers shows that a measurement run can easily lead to several GBs of data.

The CCD-cameras used within the scope of this thesis were

- a Princeton Instruments PI-LCX: 1,300 courtesy of Gerhard Grübel's group at HASYLAB, Hamburg. This camera has $1,340 \times 1,300$ pixels with a pixel size of $20 \times 20 \, \mu m^2$.
- an Andor iKon-M camera provided by the beamline ID10A at the ESRF itself. It has $1,024 \times 1,024$ pixels with a pixel size of $13 \times 13 \, \mu m^2$.

These cameras were either controlled via Roper Scientific's WinView program or directly via spec, the software operating the beamline. Depending on the program used for controlling the camera the data files use different formats:

- If controlled by spec, the data files use the .edf-format. This stands for ESRF Data Format. It has an ASCII-header of 1,024 bytes giving details about the data, after which the data are stored by the number of ADU (analog-to-digital units) for each pixel, formatted in the machine's native unsigned two-byte integer format.
- If controlled by WinView, the data files use the .spe-format. Here the header length is 4,100 bytes and it is not human-readable, after that the data are saved as with .edf-files.

Principally the two formats are equivalent if one does not read out the header: after accounting for the different header lengths the same procedures can be used for reading the data. The version of WinView used for the experiments, however, normally would not write the frames immediately after reading out the camera, it would rather read all the data into the memory and only write it to the hard disk after the measurement run is finished. This is unacceptable due to memory restrictions and the possibility of crashes, fortunately Lorenz-Mathias Stadler succeeded after consulting the manufacturers in installing a patch and writing a macro to allow the immediate write-out.

The principle of a CCD-camera is that the incident photons excite electrons via the inner photoelectric effect, these electrons are then read out and counted. For XPCS direct-illumination CCDs are used, this means that the X-ray photons fall directly onto the chip as opposed to being converted to optical photons via a phosphor screen. Charge leakage, however, leads to the so-called dark current, i.e. a charge is accumulated in the pixels and read out even in the absence of impinged X-ray photons.

Figure 5.1 shows a 100×100 detail of a frame taken with the Andor CCD in a pseudo-color display. The droplets of charge corresponding to single detected

Fig. 5.1 Detail of a weakly illuminated frame

photons are clearly discernible, only in the upper right corner there are two very bright pixels where several photons have impinged. This is a frame fulfilling the criterion of weak illumination, i.e. most of the pixels have no photon impinged onto them.

5.2 Subtracting the Dark Current

For detecting photons in the generated charge in the CCD-chip first the background has to be subtracted. Conventionally a number of dark frames (i.e. frames taken with closed shutter) is taken, the average of which is saved as the dark file and subtracted from each illuminated frame.

This thesis deals with diffuse scattered intensity, however, which leads to the data frames being only weakly illuminated, therefore most of the pixels in a frame measure only dark current also during the measurement. This fact was used for obviating the need for dark frames by obtaining the background from the illuminated files: A number of frames are read into memory, the number is chosen with respect to the amount of memory available. Then for each pixel the median of the ADUs in this pixel over all loaded frames is determined and the histogram of the ADUs in this pixel, centred on the median, is computed, with one bin for each discrete value of ADUs. The number of bins in this histogram needs to be only on the order of the standard deviation of the dark current in this pixel, which is estimated first and can be ascertained and, if necessary, adjusted afterwards. Two additional bins hold the

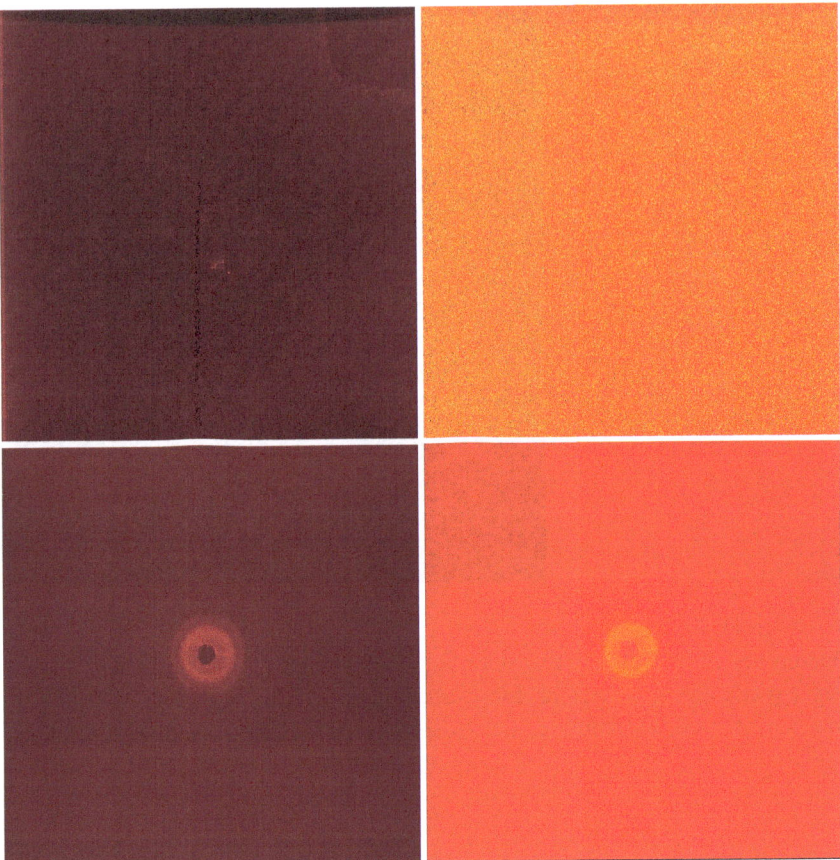

Fig. 5.2 Dark files for both the Andor CCD (*above*) and the PI CCD (*below*), in each case mean values (*left*) and standard deviations (*right*). The pseudo-colors are autoscaled, for values see text

values too high or too low to be added to the histogram. Then the frames left (if there are any) are read and added to the histograms. Finally for each pixel the highest and lowest N values are discarded and the mean value and the standard deviation of the rest is computed, where N is a value chosen beforehand. A sensible choice for N would be about 15% of the overall number of frames. If this recipe is not possible because the number in either the "too low"- or the "too high"-bin is greater than N, N is increased only for this pixel, but for sensible choices of N this can only happen if the pixel is malfunctioning. The mean values and standard deviations are written to `darkmean.bin` and `darkstd.bin`, respectively, in single precision 4-byte floating-point format, such that they can be viewed conveniently via `fit2d`, ESRF's standard 2d-file viewer.

The rationale for neglecting the highest N values is that the value for the dark current should not be influenced by the charges due to X-rays, the lowest N values are neglected for symmetry. As only a small number of pixels actually has additional

Fig. 5.3 Distribution of the deviations of the actual ADU values from the mean for the Andor CCD, fitted by a Gaussian distribution. The histogram is computed with bins from −40 to 40 ADUs, values outside of this interval are assigned to the extreme bins

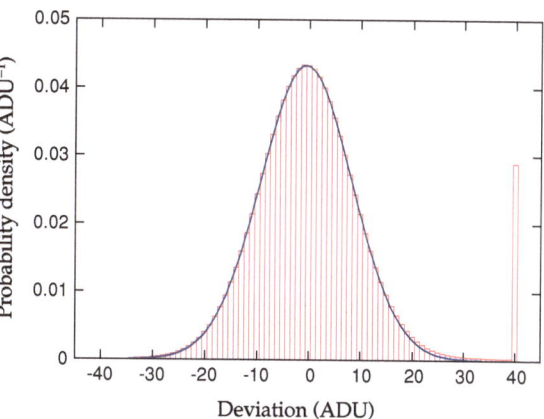

charge, the bias towards too high mean values should be small with this course of action.

Dark files for both cameras are given in Fig. 5.2. Both CCDs were very new and therefore in very good condition with no destroyed pixels, but in previous small-angle scattering experiments they obviously had already been over-exposed, resulting in local damages to the silicon chip and therefore increased dark current. This is per se not a problem, as the dark current is subtracted anyway, but in the case of the PI CCD also the variance of the dark current is locally increased.

The homogeneous dark-red area of the panel pertaining to the Andor's mean dark values corresponds to 3,436 ADUs, with a pixel-to-pixel standard deviation of these mean values of only 0.7 ADUs. The discernible arc at the very top (probably due to the fabrication of the chip) is three ADUs less. The standard deviations of the dark current in a pixel from frame to frame after removing the highest and lowest values (Fig. 5.2 right top) is 5.5 ADUs with no significant pixel-to-pixel variations, apart from the barely discernible arc, there it is 5.4 ADUs. These values for the PI chip are a mean value of 319 ADUs with a pixel-to-pixel variation of five ADUs, the values in the central ring are on the average 367 ADUs. The standard deviations are 3.6 ADUs, in the ring 4.4 ADUs.

Figure 5.3 shows the distribution of the actual read-out ADU values relative to the mean computed as stated above. Apart from the about 3% of pixels which have more than 40 ADUs additional charge due to detected X-rays, the distribution is Gaussian, with a fitted width of 8.9 ADUs and a centre of −0.5 ADUs. The width is higher than the above-quoted 5.5 ADUs because the tails are not cut here (apart from the 3% due to X-rays), and the discrepancy in the mean is due to asymmetry induced by the detected X-rays. These −0.5 ADUs per pixel are a systematic error introduced with this method of computing the dark current, but with a photon corresponding to about 2,000 ADUs as shown in the next section, it is made up for by the advantages of this method (a very good statistical accuracy without having to spend time taking dark frames).

The leakage current is not absolutely constant over the duration of a measurement, for example for the measurement evaluated above it rose by about two ADUs over the measurement duration of 2 h. This is most probably due to drifts in the temperature of the CCD chip. The programs can take such drifts into account by modifying the dark values dynamically, but a drift of two ADUs is not relevant given the amount of charge generated by an X-ray photon, so this feature was turned off.

5.3 Histogram of the Droplet Charges

At certain times it is desirable to generate a histogram of the droplet charge distribution. This includes the start of the experiment (for learning the characteristics of the camera) and after changing the sample or the set-up (in order to check for fluorescence coming from the sample or from the primary beam hitting parts of the equipment).

From Fig. 5.1 it is obvious that the droplets have a variety of shapes: This is due to the fact that the photoelectric absorption of a photon generates a cloud of charges on the chip, which can be assumed Gaussian and whose width is on the order of the CCD's pixel size (see Miyata et al. [2]). Depending on the centre of the charge cloud relative to the pixel boundaries, this can lead to a droplet comprising only one pixel (when the centre of the charge cloud is near the centre of a pixel), two pixels (when the charge cloud intersects the boundary between pixels along one dimension), or four pixels (when the centre of the charge cloud is near a pixel corner). In the last case the fourth pixel's charge can be lost in the fluctuations of the dark current if it is very weak, leading to three-pixel droplets. Apart from these single-photon droplets a small fraction of droplets comes from overlapping charge clouds, leading to larger droplets of arbitrary shape.

In the scope of this thesis droplets were treated in a model-free approach: a droplet is defined as a connected set of pixels (where the neighbours of a given pixel are the four nearest neighbours in the square lattice), each with a charge significantly above the mean dark current of the respective pixels. This is implemented in `hist` and `evaluatedroplets` in the following way: The programs use a data frame, a logical array of the same size as the data frame initialized to `false`, and a stack which is initially empty. First the dark file is subtracted from the data frame. Then the pixels are gone through and tested one after the other if both their logical state is `false` and if the value in the pixel is significantly higher than zero. If this is the case, a counter is initialized to zero, the pixel's logical state is set to true, and its index is pushed into the previously empty stack. Then the operation turns to the stack: the uppermost index is taken and the corresponding pixel's value is added to the counter. Then the four neighbours of the pixel are tested sequentially on their logical state and their value as before, and for a positive result their state is set to `true` and their index pushed into the stack. Then the next index is popped from the stack, until the stack is empty. The value in the counter is then the sum of the values of the pixels in the droplet and the function returns to the loop over the frame's pixels.

Figure 5.4 shows histograms of droplet charges calculated for the Andor and the PI chips. The peaks corresponding to an integral number of elastically scattered photons are clearly discernible. The plots with the linear scale show that the majority

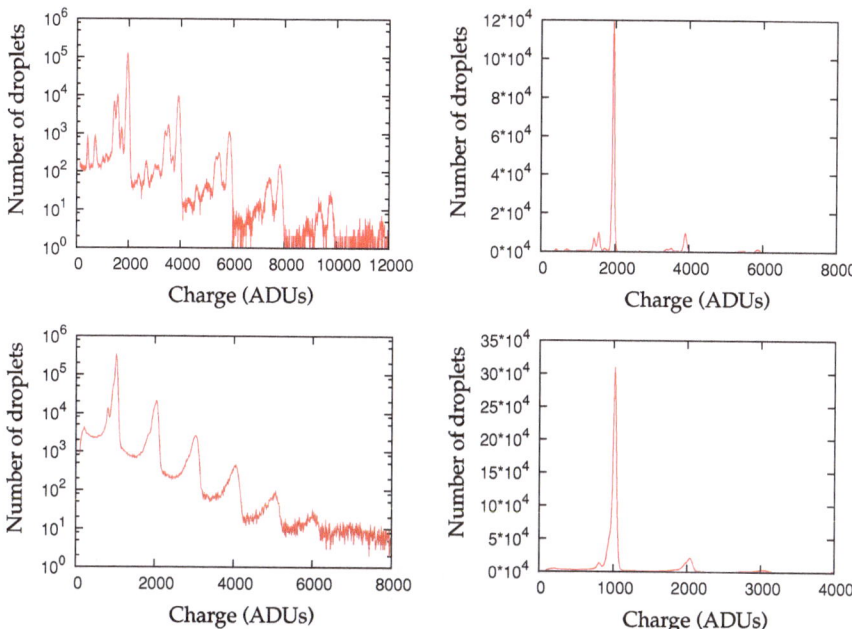

Fig. 5.4 Histograms of droplets for both the Andor CCD (*above*) and the PI CCD (*below*), in each case on a logarithmic (*left*) and a linear scale (*right*). Bin width is 10 ADUs, incident photon energy 8 keV

of droplets correspond to one photon. The number of ADUs per photon (the position of the one-photon peak) for the Andor CCD is 1,955, for the PI CCD it is 1,018. Even though the ratio of the variations of the dark current to the charge per photon is approximately the same for both chips, the spectrum detected by the Andor CCD shows far more details than the spectrum of the PI chip. This difference is most likely due to the architecture of the chip, probably the PI CCD is not so efficient in collecting all the generated charges if the photon was absorbed deep in the chip. The features visible in the Andor's spectrum can principally come from two distinct processes: on the one hand they can be fluorescence lines from elements in the sample or in the furnace, on the other hand they can be so-called escape peaks, when the energy of a photon is not completely transferred to electronic excitations in the chip, but instead a silicon atom is excited and a fluorescence photon is emitted. The following table gives the positions and the estimated weights relative to the main elastic line at 8.0 keV with the processes possibly responsible for them. The X-ray energies are taken from the X-ray data booklet [4].

Now also the limits for considering a droplet as pertaining to a single elastically scattered photon can be set. The linear plots in Fig. 5.4 show that for the Andor CCD the limits can safely be set to 1,800 and 2,100 ADUs. For the PI CCD this decision is not so clear, here the limits were set to 920 and 1,110 ADUs.

Position (keV)	Weight (relative)	Possible source
8.0	1	One-photon elastic line
6.42	0.083	FeKα (6.40 keV), MnKβ (6.49 keV), SiKα-escape (8.0–1.74 keV)
5.91	0.057	MnKα (5.90 keV)
7.10	0.012	FeKβ (7.06 keV)
2.91	0.0071	?
1.73	0.0035	SiKα (1.74 keV)

Fig. 5.5 Path of a cosmic ray

5.4 Detecting Photon Events

The next step in evaluating the data is extracting the positions of the absorbed photons from the frames with `evaluatedroplets`. The droplets are detected and evaluated with the same algorithm as in the previous section, but now not only the sum of the pixels' ADUs is computed, but also the moments in both x- and y-direction. Only the droplets with a number of ADUs within the limits set for a single elastically scattered photon are further considered, the reason is that contrary to the claims of Livet et al. [1] it is not so trivial to determine the exact positions of the incidence of single photons within a multi-photon droplet. Not considering these photons is conservative as it can only decrease the contrast, also only a very small number of photons is lost in this way, see Fig. 5.4. This also does away with the occasional cosmic ray, see Fig. 5.5.

The *x*- and *y*-moments of the one-photon droplets are sequentially written to memory, simultaneously an array of the summed counts for each pixel over the whole measurement (i.e. the moments of the droplets are rounded to integers and assigned to a pixel) and a vector of the number of photons for each frame is generated. Using the array of the summed photons per pixel the data are rearranged so that they are described by the frame numbers when photons were detected in a given pixel. These numbers are written to data.bin, sequentially for all pixels, the fractional parts of the moments are written to datafrac.bin. Finally the array of the summed photons is written to summedframes.bin as binary short integers (can be viewed in fit2d) and the vector of the photons per frame is written to timeseries.bin as binary integers and to timeseries.txt in ASCII-form (for plotting).

When evaluating a measurement run generally the first thing is to compute the dark file, then evaluatedroplets is run. For doing that it is advisable to read in the frames starting from the last one in order to exploit caching, which is done both by the hard-disk and by the operating system. This considerably speeds up the process, particularly if the data are on an external hard-disk with the concomitant low transfer rates.

5.5 Computing the Auto-Correlation Function

After having detected the photons the auto-correlation function can be computed. This is done by the program computeacf. The real situation at a synchrotron is more complicated than as was considered in Sect. 3.6, however: in general the intensity of the incoming beam is not constant, because first the ring current decays between the refills[1] and second the position of the beam moves over the slits on the timescale of hours. Neglecting to consider these effects would result in the auto-correlation function's being dominated by the fluctuations of the incident radiation as opposed to measuring the fluctuations coming from the sample. Fortunately the number of pixels on the detector is large, so the intensity in a pixel at a given time can be normalized by the instantaneous expected value of the intensity, given by the average intensity of all pixels at this time. Such an approach is valid because the expected value of the auto-correlation function at non-zero time delay is not affected by quantization as was shown in Sect. 3.6.

Thanks to the rearranging described above the computation can be done very efficiently. If the times when a photon was detected in a given pixel are denoted by $t_1 \ldots t_M$ and the number of photons in frame i by N_i for all i, then the auto-correlation function is obtained by iterating over all $1 \leq i < j \leq M$ and adding $1/(N_{t_i} N_{t_j})$ to the entry corresponding to the value of the auto-correlation function for the time delay $t_j - t_i$. This is iterated for all pixels. The entry corresponding to the time delay

---

[1] Modern synchrotrons starting with the APS operate in the so-called top-up mode, meaning that the refilling of the electrons happens practically continuously. This is also planned to be implemented with the ESRF upgrade.

Fig. 5.6 Two-time
auto-correlation function of
262 frames in pseudo-color
display, smoothed by a
Gaussian kernel with a width
of three frames. The left top
corresponds to
$(t_1 = 0, t_2 = 0)$, the colour
of the off-diagonal areas to a
value of 1 (i.e. no
correlation), the average
yellow of the diagonal to
1.13 (i.e. a contrast of 0.13)

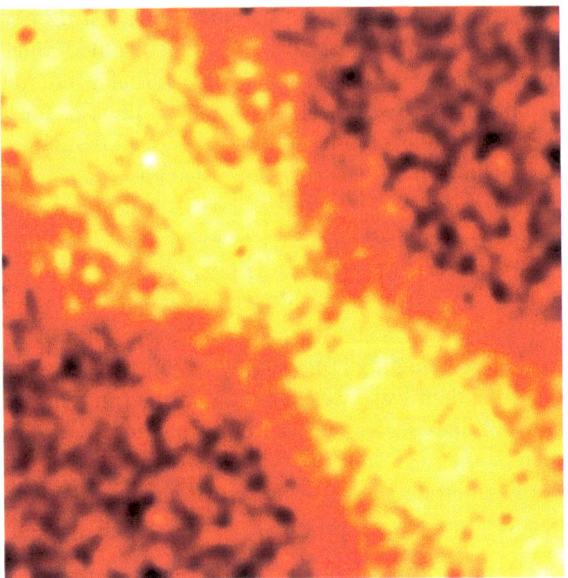

Δt is then multiplied by $(T - \Delta t)/K$ where T is the number of frames and K the number of pixels, and the resulting values are written to a text file.

The two-time auto-correlation function, which is defined in analogy to Eq. 3.6.2 but without averaging over absolute time

$$g^{(2)}(\boldsymbol{q}, t_1, t_2) = \frac{\langle I(\boldsymbol{q}, t_1) I(\boldsymbol{q}, t_2) \rangle}{\langle I(\boldsymbol{q}, .) \rangle^2}, \tag{5.5.1}$$

can be computed equivalently by adding up the entries of a matrix describing absolute times as opposed to a vector describing time delays. The results are written to a file as 4-byte floats suitable for viewing by fit2d. This thesis deals with equilibrium dynamics, so the two-time auto-correlation function was only evaluated qualitatively in order to ensure that the sample was truly in equilibrium. Also sudden correlation losses can happen, probably due to instabilities of the set-up, an example of which is given in Fig. 5.6: homogeneous dynamics should give a band of constant width along the diagonal.

Contrary to what one would naïvely think, the fact that the charges generated by a photon are in most cases distributed over a number of pixels (described by the so-called point-spread function) is not undesirable because it potentially gives sub-pixel resolution as demonstrated by Miyata et al. [3]. Unfortunately, the relation between the centre of the cloud of charges generated by the photon and the moments in x- and y-direction of the ADUs in the droplet is a nonlinear function, which depends on the width of the charge cloud and the pixel size. The width of the charge cloud is given by the distances the charges had to diffuse to the electrode [2], and it therefore depends on how deep within the chip the photon was absorbed. This depth

Fig. 5.7 Illustration of the splitting into subpixel. The physical pixel is dashed, the four distinct translations (displaced by a small amount for visual clarity) defining the logical pixels are solid

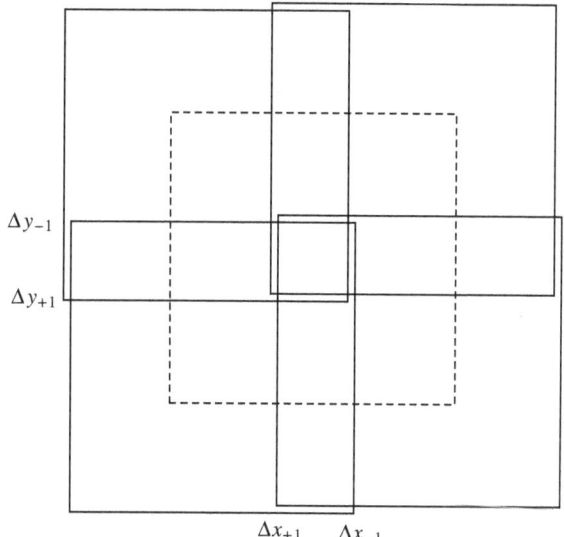

Δy_{-1}

Δy_{+1}

Δx_{+1} Δx_{-1}

is a random variable for each photon, so it is principally not possible to assign the exact position of detection to each droplet. A rough estimate on the position within the pixel, that means an assignment to logical sub-pixels, can be made, however. Back-illuminated CCDs with their longer diffusion length, therefore larger charge cloud and in turn less severe effects of discretization would be highly desirable for these kind of experiments.

In the course of this thesis the fractional parts of the moments in datafrac.bin were considered in three ways:

- *The straight-forward way*: The logical pixels used in the algorithm above are identical to the physical pixels of the CCD. The fractional parts of the moments are consequently not considered at all.
- *More involved but still conservative*: The grid of the logical pixels is translated relative to the physical pixels by a small amount and the results are averaged over a number of such translations.
- *The optimistic way*: The photons in a physical pixel are reclassified into several logical sub-pixels.

The first approach was taken for the experiments with the PI CCD, first because the more intricate algorithms had not been implemented then, second due to the poorer performance as visible in the histograms trying to achieve sub-pixel resolution did not seem so promising for this CCD.

The second approach was implemented in the following way: a histogram of the fractional parts of the moments in the x-dimension $p(\Delta x)$ was prepared, the centre of this distribution Δx_0 was computed via the complex angle of its first Fourier component, and Δx_{-1} and Δx_{+1} were chosen such that both the integrals over

$[\Delta x_{-1}, \Delta x_0]$ and $[\Delta x_0, \Delta x_{+1}]$ were equal to 0.25. The same was done for the y-dimension, and the translations used for averaging were the four possible choices of $(\Delta x_{\pm 1}, \Delta y_{\pm 1})$, see Fig. 5.7. As the classification with respect to these values Δx_i and Δy_i should not be much worse than the straight-forward classification (i.e. with a splitting fractional value of 0.5), the obtained auto-correlation function should have the same statistical accuracy and essentially the same contrast. However, averaging over the four distinct translations improves the statistical accuracy of the result: disregarding the denominator the value of the auto-correlation function is given by the number of incidences N when in the same logical pixel photons were detected with a given time delay. Inspecting Fig. 5.7 it can be seen that if both photons were detected in the same quarter-subpixel then they are counted in all four translations; if they were in neighbouring subpixel, then in two translations; and if their subpixels only share a corner, then they are counted only in one translation. The entry in the auto-correlation function after averaging N is therefore given by

$$N = \frac{1}{16}\left(4N_{00} + 2N_{01} + 2N_{0\bar{1}} + 2N_{10} + 2N_{\bar{1}0} + N_{11} + N_{1\bar{1}} + N_{\bar{1}1} + N_{\bar{1}\bar{1}}\right)$$

$$(5.5.2)$$

where N_{00} is the number of incidences in the same subpixel, N_{10} the number of incidences with the second photon in the right neighbouring subpixel of the first and so on. For exclusively Poisson noise (i.e. for low contrast) these numbers are uncorrelated and they have the same variance v, so the variance of N is given by

$$V(N) = \frac{v}{16^2}\left(16 + 4 \cdot 4 + 4 \cdot 1\right) = \frac{9v}{64}.$$

$$(5.5.3)$$

Calculating the auto-correlation function with only one choice of logical pixels would correspond to

$$N' = \frac{1}{4}\left(N_{00} + N_{01} + N_{10} + N_{11}\right)$$

$$(5.5.4)$$

with a variance

$$V(N') = \frac{v}{4}.$$

$$(5.5.5)$$

The ratio of $V(N)/V(N')$ is therefore 9/16 which shows that using this approach lowers the standard deviations of the auto-correlation function in the ideal case by a factor of 0.75 with respect to the straight-forward approach.

For the third approach the quarter-subpixels are directly used as logical pixels. Conducting an experiment with half the detector distance would then give the same situation as using the physical pixels with the original distance (provided the count rate is low enough, i.e. the single droplets can still be resolved), but with the fourfold number of logical pixels, thereby lowering the standard deviations of the points of the auto-correlation function by a factor of 0.5.

In principle one could also use a finer logical resolution given the necessary low count rates. The problem is, however, that there is a considerable fraction of droplets which consist of only one pixel. The size of the subpixels is obviously limited from below by the fact that these droplets all have to go into one subpixel. This is also the reason for choosing the limits between the subpixels Δx_i and Δy_i the way it was done: The classification can be done best if the charge is spread approximately equally between two pixels. The fact that the photons are absorbed in different depths in the chip and that therefore the charge clouds have different sizes also leads to the two-dimensional distribution of fractional moments $p(\Delta x, \Delta y)$ not factorizing into $p(\Delta x)p(\Delta y)$. Therefore the numbers of counts in the quarter-subpixels defined as above is not equal, rather the numbers of the subpixels corresponding to the centres of the physical pixels and the subpixels corresponding to the corners of the physical pixels are higher than the numbers of the subpixels corresponding to the edges. This does not invalidate the second approach, only the factor in Eq. 5.5.3 will get a bit higher. The third approach, however, needs a classification such that all the logical pixels have the same expected value of counts, otherwise an artificial apparent contrast is introduced. Therefore a more complicated definition of subpixels would be needed. This has not been implemented yet, the data obtained by the Andor CCD were therefore evaluated by the second approach.

5.6 Fitting the Auto-Correlation Functions

For quantitative analyses the auto-correlation function as written by `computeacf` into acf.txt has to be fitted by a function of a form as predicted by the theory. In the most simple case this is just 1 plus an exponential with fitted decay time and fitted coherence factor

$$g^{(2)}(\Delta t) = 1 + \beta e^{-\Delta t/(2\tau)}, \qquad (5.6.1)$$

obtained by plugging Eq. 2.3.19 into Eq. 3.6.12.

Fitting was done with the program gnuplot, as this is a small, fast, flexible, and free tool, running on all relevant platforms. If the number of data points in the auto-correlation function is T then the variance of the data point corresponding to Δt is inversely proportional to $T - \Delta t$, because this is the number of pairs of frames which can be correlated. Therefore the weight given to each data point in fitting should be proportional to $T - \Delta t$. For gnuplot the weight has to be specified via a quantity proportional to the expected standard deviation, that is $\sqrt{T - \Delta t}$. gnuplot does not seem to give the user the ability to access the length of the data set explicitly, but the following algorithm, where acf is a string holding the path and the filename of the auto-correlation function, does the trick:

```
fit T acf u 0:0 via T
T=ceil(T*2+.5)

g(x)=beta*exp(-2*x/tau)+1
fit [x=1:*] g(x) acf u 0:1:(1/(sqrt(T-$0))) via beta,tau
```

The fit range has to be restricted to positive time delays as the value of the autocorrelation function for time delay $\Delta t = 0$ is dominated by the influence of Poisson noise.

References

1. F. Livet, F. Bley, J. Mainville, R. Caudron, S.G.J. Mochrie, E. Geissler, G. Dolino, D. Abernathy, G. Grübel, M. Sutton, Using direct illumination CCDs as high-resolution area detectors for X-ray scattering. Nucl. Instrum. Methods A **451**, 596 (2000)
2. E. Miyata, M. Miki, J. Hiraga, D. Kamiyama, H. Kouno, H. Tsunemi, K. Miyaguchi, K. Yamamoto, Mesh experiment for back-illuminated CCDs in improvement of position resolution. Nucl. Instrum. Methods A **513**, 322 (2003)
3. E. Miyata, M. Miki, H. Tsunemi, J. Hiraga, H. Kouno, K. Miyaguchi, (2002) Direct x-ray imaging of μm precision using back-illuminated charge-coupled device. Jpn. J. Appl. Phys. **41**, L500 (2002)
4. A. Thompson et al., X-ray data booklet (Lawrence Berkeley National Laboratory, 2001), http://xdb.lbl.gov

Chapter 6
Considerations Concerning the Experiment

In this chapter I will present some useful calculations for setting up the experiment. First the signal-to-noise ratio (the inverse of the relative standard error of the fitted correlation time) is computed as a function of the number of counts per pixel, the coherence factor, and the correlation time, leading to the identification of the product of intensity and coherence factor as the figure-of-merit. In the second section the optimization of the beamline set-up with regard to this figure-of-merit is discussed.

6.1 Counting Noise

The statistical errors in an XPCS measurement have two sources: the stochastic nature of the fluctuations in the sample (and therefore of the scattered radiation) and the stochastic process of photon quantization. In wide-angle scattering it is justifiable to treat all the approximately 10^6 pixels on the detector as equivalent within the experimental accuracy, i.e. belonging to the same q. The measured auto-correlation function then samples over 10^6 distinct evolutions of the intensity, and the mean evolution has therefore a relative accuracy on the order of 10^{-3}. The number of scattered photons in the diffuse regime, however, is low with today's X-ray sources, even with samples selected for their scattering efficiency. Therefore the attainable accuracy is governed by the Poisson noise of photon quantization. In this section I will calculate the statistical accuracy of the correlation time fitted onto the measured auto-correlation function as a function of the actual correlation time and the experimental parameters.

It comes in handy here to measure time by frames. Let $G^{(2)}(k)$ be the experimental auto-correlation function before normalization, i.e. it is the product of the number of photons detected a time interval of k frames apart, averaged over all pixels and absolute time. Let N be the number of pixels and K the number of frames. Let $p(I_1, I_2)$ be the joint probability density of the squared modulus of the electrical field (i.e. the intensity before quantization) at two times which are k frames apart. As already postulated in Sect. 3.6 the expected value of $G^{(2)}(k)$ with $k > 0$ is given by

M. Leitner, *Studying Atomic Dynamics with Coherent X-rays*,
Springer Theses, DOI: 10.1007/978-3-642-24121-5_6,
© Springer-Verlag Berlin Heidelberg 2012

$$E\left(G^{(2)}(k)\right) = \int dI_1 dI_2 p(I_1, I_2) \sum_{n_1} \frac{I_1^{n_1}}{n_1!} e^{-I_1} \sum_{n_2} \frac{I_2^{n_2}}{n_2!} e^{-I_2} n_1 n_2$$

$$= \int dI_1 dI_2 p(I_1, I_2) I_1 I_2. \tag{6.1.1}$$

The variance of $G^{(2)}(k)$ is given by

$$V\left(G^{(2)}(k)\right) = \frac{1}{N(K-k)} \left(\int dI_1 dI_2 p(I_1, I_2) \sum_{n_1} \frac{I_1^{n_1}}{n_1!} e^{-I_1} \sum_{n_2} \frac{I_2^{n_2}}{n_2!} e^{-I_2} n_1^2 n_2^2 - E\left(G^{(2)}(k)\right)^2 \right)$$

$$= \frac{1}{N(K-k)} \left(\int dI_1 dI_2 p(I_1, I_2)(I_1 + I_1^2)(I_2 + I_2^2) - E\left(G^{(2)}(k)\right)^2 \right) \tag{6.1.2}$$

For low count rates, that is for $E(I_j) \ll 1$, only the term of lowest order in I_j contributes, leading to

$$V\left(G^{(2)}(k)\right) = \frac{1}{N(K-k)} E\left(G^{(2)}(k)\right). \tag{6.1.3}$$

The reason for the statistical inaccuracy to rise with the length of the time interval k is that the number of pairs of frames which are k frames apart falls, in the extreme case where $k = K - 1$ there is only the pair consisting of the first and last frame left.

Due to Eq. 3.6.2 the normalization for the auto-correlation function is to divide by the square of the mean intensity

$$g^{(2)}(k) = \frac{G^{(2)}(k)}{\bar{I}^2}, \tag{6.1.4}$$

therefore the variance of the normalized auto-correlation function is given by

$$V\left(g^{(2)}(k)\right) = \frac{E\left(g^{(2)}(k)\right)}{\bar{I}^2 N(K-k)}. \tag{6.1.5}$$

Here I neglected the uncertainty of the measured mean intensity $\bar{I}$, which is smaller than the uncertainty of the numerator by orders of magnitude for small intensities. Plugging in an exponential decay leads to the intermediate result

$$V\left(g^{(2)}(k)\right) = \frac{1 + \beta e^{-2\Gamma k}}{\bar{I}^2 N(K-k)}. \tag{6.1.6}$$

In this section I write Γ for the inverse of the correlation time τ for notational convenience.

What is more interesting than the statistical accuracy of one point in the experimental auto-correlation function is the fitted decay time (or its inverse Γ). This is a non-linear problem, as the fitted auto-correlation function is a non-linear function of Γ. It can be linearized, however, and fortunately the linearized problem is equivalent to the non-linear problem for small uncertainties.

Suppose that the non-linear least-squares problem is solved by $\hat{g}(k; \hat{\beta}, \hat{\Gamma})$, yielding the fitted parameters $\hat{\beta}$ and $\hat{\Gamma}$. The idea is now to pretend to fit the residuals by a linear combination of the partial derivatives of $\hat{g}$ with respect to the parameters. It is clear that the best fit to the residuals is with the coefficients of both partial derivatives equal to zero, because otherwise $\hat{\beta}$ and $\hat{\Gamma}$ would not solve the original problem. The new problem, however, is linear, allowing standard techniques for estimation of the parameter uncertainty to be employed:

Let $\mathbf{X}$ be the matrix of the basis functions of the linear problem to be fitted, i.e. in this case

$$
\mathbf{X} = \begin{pmatrix} \vdots & \vdots \\ \frac{\partial}{\partial \beta} \hat{g}(k; \beta, \Gamma)|_{\hat{\beta}, \hat{\Gamma}} & \frac{\partial}{\partial \Gamma} \hat{g}(k; \beta, \Gamma)|_{\hat{\beta}, \hat{\Gamma}} \\ \vdots & \vdots \end{pmatrix} = \begin{pmatrix} \vdots & \vdots \\ e^{-2\Gamma k \Delta t} & -2k \Delta t \beta e^{-2\Gamma k \Delta t} \\ \vdots & \vdots \end{pmatrix}
$$
(6.1.7)

where k runs from 0 to $K - 1$ and Δt is the temporal spacing of the frames. Let $\mathbf{M}$ be the covariance matrix of the entries of the experimental auto-correlation function, that is

$$
M_{k,j} = \delta_{k,j} \frac{1 + \beta e^{-2\Gamma k}}{\bar{I}^2 N(K - k)}
$$
(6.1.8)

due to Eq. 6.1.6 and the fact that entries corresponding to distinct times are uncorrelated in the limit of small $\bar{I}$.

The covariance matrix of the fitted parameters is then given by

$$
\hat{\mathbf{M}} = \left(\mathbf{X}^* \mathbf{M}^{-1} \mathbf{X} \right)^{-1},
$$
(6.1.9)

with the variance of the inverse of the fitted correlation time in its lower right entry. The signal-to-noise ratio of the fitted correlation time is then given by

$$
\frac{E(\hat{\tau})}{\sqrt{V(\hat{\tau})}} = \frac{E(\hat{\Gamma})}{\sqrt{V(\hat{\Gamma})}} = \sqrt{N} I_0 \beta f(\tau/T, \beta, K)
$$
(6.1.10)

where T is the duration of the experiment, $I_0 = K \bar{I}$ are the mean counts per pixel over the whole experiment, and $f(\tau/T, \beta, K)$ has a finite limit for both $\beta \to 0$ or $K \to \infty$.

I want to elaborate on this result a bit. For a fixed correlation time with respect to the duration of the experiment the signal-to-noise ratio goes with the square root of the number of pixels averaged over, which is immediately obvious. The linearity of the dependence on the coherence factor β in first order is plausible as the relative uncertainty of the points of the auto-correlation function decreases linearly with β. The linearity in I_0 is probably not so obvious at first glance, but as for low I_0 most of the frames have no photon in a given pixel, only a small number has one photon, and higher counts can be neglected, the auto-correlation function essentially just counts

Fig. 6.1 Influence of the ratio between the correlation time τ and the duration of the experiment T on the signal-to-noise ratio. The upper blue curve is for the limit of small β as given in Eq. 6.1.11, the middle red curve and the lower green curve are for $\beta = 0.1$ and $\beta = 1$, respectively, obtained by numerically solving Eq. 6.1.9 with large K

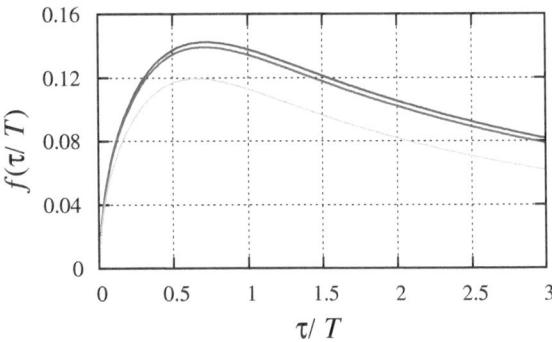

the pairs where there was one photon in both frames. This number is quadratic in I_0, its standard deviation is therefore linear, and its relative standard deviation goes with I_0^{-1}.

The behaviour of the signal-to-noise ratio in its general form is best evaluated numerically. Doing the summations the experimentally relevant limiting case of small β and large K can be obtained analytically with reasonable effort, however:

$$f(\tau/T, \beta \to 0, K \to \infty) = \frac{1}{2x} \sqrt{\frac{e^{-x}x^3 + (e^{-x} + 1)x^2 + 4(e^{-x} - 1)x + 2(e^{-x} - 1)^2}{x + e^{-x} - 1}}$$
(6.1.11)

where x is shorthand for $4T/\tau$.

For the behaviour of $f(x)$ consult Fig. 6.1. The increase with $\sqrt{\tau/T}$ for small correlation times is due to the increasing number of significant (i.e. larger than 1) points in the auto-correlation function, the decrease with T/τ for long correlation times is due to the decrease in the observable magnitude of the decay (i.e. the amount by which the auto-correlation function has decayed before the experiment is over).

The optimal choice of the correlation time is therefore a value τ/T of about 0.7, such that the auto-correlation function $1 + \beta e^{-2\Delta t/\tau}$ shows a decay to about 6% of its initial deviation from 1. This only holds, however, if one has absolute confidence in the stability of the set-up and the beam. In realistic situations it is very reassuring to choose somewhat smaller correlation times so as to be able to compute the two-time auto-correlation function and confirm visually that what one measures are indeed equilibrium dynamics and not artefacts, in the worst case excluding such artefacts from the evaluation.

6.2 Optics and Contrast

In Sect. 3.6 the influence of the dynamics in the sample on the intensity-intensity auto-correlation function was treated. This was done under the assumption of ideal circumstances, i.e. the incoming radiation was considered to be an ideal monochromatic plane wave, and the scattered radiation to be detected in a point. Here I will address the non-ideal case, that is to compute the coherence factor β of Eq. 3.6.12 for a given experimental set-up. As a concrete example relevant to this thesis I will use the values pertaining to the beamline ID10A at the ESRF.

In a synchrotron the incident radiation is generated by the relativistic motion of the electrons through the undulator. The electrons in the storage ring are independent, therefore the radiation generated by one electron has no phase relation with the radiation generated by another electron.[1] The intensity in a given pixel can therefore be thought of as the sum of the intensities due to distinct electrons, detected at distinct points within the detector's pixel. For a given point in time let $I_1 = A_1 A_1^*$ be such a contribution with

$$A_1 = \int \mathrm{d}\vec{x}\, A_1(\vec{x})\rho(\vec{x}), \tag{6.2.1}$$

that is $A_1(\vec{x})$ is the amplitude with which an electron located at $\vec{x}$ contributes to this specific scattering event. Apart from the normalization factor and neglecting quantization the expected value of the measured auto-correlation function at zero time delay is then given by

$$
\begin{aligned}
g^{(2)}(0) &\propto \lang\!\langle A_1 A_1^* A_2 A_2^* \rangle\!\rangle \\
&= \left\langle\!\!\left\langle \int \mathrm{d}\vec{x}_1 \mathrm{d}\vec{x}_2 \mathrm{d}\vec{x}_3 \mathrm{d}\vec{x}_4\, A_1(\vec{x}_1) A_1^*(\vec{x}_2) A_2(\vec{x}_3) A_2^*(\vec{x}_4) \left\langle \rho(\vec{x}_1)\rho(\vec{x}_2)\rho(\vec{x}_3)\rho(\vec{x}_4)\right\rangle \right\rangle\!\!\right\rangle,
\end{aligned}
\tag{6.2.2}
$$

where the inner angular brackets denote the expected value regarding the scatterer density of the sample $\rho(\vec{x})$ and the outer brackets the expected value regarding the amplitudes $A_1(\vec{x})$ and $A_2(\vec{x})$, respectively. This is nothing else than Eq. 3.6.3 at time delay 0 generalized for non-ideal circumstances. As it was done there, the way to proceed now is to use the fact that the correlations in the sample are short-range. For "nearly" ideal circumstances the sample correlation length ξ is much shorter than the correlation lengths of the radiation, that is the length over which differences between the amplitudes $A_1(\vec{x})$ and $A_2(\vec{x})$ emerge (apart from a trivial constant phase offset), so it suffices to consider the sample correlation function as

$$\left\langle \rho(\vec{x}_1)\rho(\vec{x}_2) \right\rangle \propto \delta(\vec{x}_1 - \vec{x}_2) + c. \tag{6.2.3}$$

By factorizing the four-point correlation in Eq. 6.2.2 as in Sect. 3.6 and observing that the constant vanishes from the integration[2] due to the fast fluctuation of $A_i(\vec{x})$, we arrive at

---

[1] This is the difference from X-ray sources of the fourth generation (X-ray free electron lasers), where the electric field generated by the electrons feeds back and bunches the electrons, enforcing the phase relation.

[2] In the scatterer's language the argument is that we are not in the forward direction.

$$g^{(2)}(0) \propto \left\langle \int d\vec{x}_1 A_1(\vec{x}_1) A_1^*(\vec{x}_1) \int d\vec{x}_2 A_2(\vec{x}_2) A_2^*(\vec{x}_2) \right.$$
$$\left. + \int d\vec{x}_1 A_1(\vec{x}_1) A_2^*(\vec{x}_1) \int d\vec{x}_2 A_1(\vec{x}_2) A_2^*(\vec{x}_2) \right\rangle. \qquad (6.2.4)$$

For incoherent radiation $A_1(\vec{x})$ and $A_2(\vec{x})$ are independent from each other, therefore the second term vanishes. In this case the normalized intensity-intensity auto-correlation function $g^{(2)}(\Delta t)$ has to be constant equal to one, so the correct normalization factor has to be the inverse of the first term, leading to

$$g^{(2)}(0) = 1 + \left\langle \frac{|\int d\vec{x} A_1(\vec{x}) A_2^*(\vec{x})|^2}{\int d\vec{x} A_1(\vec{x}) A_1^*(\vec{x}) \int d\vec{x} A_2(\vec{x}) A_2^*(\vec{x})} \right\rangle, \qquad (6.2.5)$$

where the second term can be equated with the coherence factor β of Eq. 3.6.12. The statistical weight of the amplitudes $A_i(\vec{x})$ obviously has to be the weight with which they contribute to the intensity on the detector.

Under simplifying assumptions Eq. 6.2.5 can in principle be evaluated analytically: Treating the incident radiation as composed of plane waves and neglecting diffraction at the beam-defining slits, the amplitude $A_i(\vec{x})$ is just the characteristic function of the illuminated volume modulated by a plane wave with wavevector $\vec{q}_i$, corresponding to the difference between the wave-vectors of the incident and the outgoing waves. The integral in the numerator of Eq. 6.2.5 is then just the Fourier transform of the characteristic function of the illuminated volume evaluated at $\Delta \vec{q} = \vec{q}_1 - \vec{q}_2$, where the probability density of these $\Delta \vec{q}$ is given by the angular size of the source and the pixel (direction) and the energy spread of the radiation (length). This simple approach works surprisingly well where it is applicable [3], but it cannot deal with focussing. This will be treated in the following.

The essential features of the optical setup at beamline ID10A at the ESRF are illustrated in Fig. 6.2. An electron passing through the undulator generates a cone of radiation concentrated around the direction of propagation of the electron. After being trimmed at the SS0-slits the wave is focussed by the CRL system (compound refractive lenses). This is just a block of Be with parabolic cavities, acting as focussing lenses, for in the X-ray regime vacuum is actually the optically denser medium. The wave then propagates to the sample, which is approximately in the focal spot. On the way it is deflected at the monochromator and immediately before the sample it is cut again by the sample slits. The scattered radiation is then detected at the detector.

The only working monochromator at ID10A is a single crystal of Si, operating in symmetric Bragg geometry at the (111)-reflection. Its theoretical reflectivity as a function of the X-ray energy [2] is shown in Fig. 6.3. The relative width of the reflectivity (FWHM) is 1.43×10^{-4}. This is much smaller than the width of the first fundamental mode of the undulator, which is on the order of a few percent, so the distribution of energies after the monochromator is given by the reflectivity of the Si crystal.

The source size is defined by the electron beam. Due to the fact that the electrons are accelerated in the horizontal direction along their way around the synchrotron,

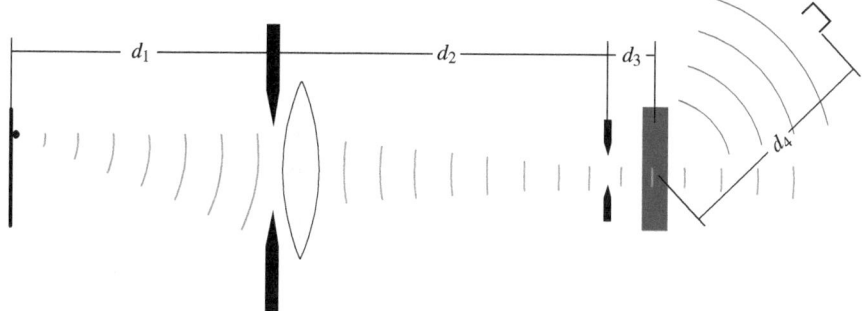

Fig. 6.2 Schematic illustration of the setup at a coherence beamline and the wave train corresponding to a given photon. The monochromator between the lenses and the sample slit is not shown

Fig. 6.3 Reflectivity of a Si-(111) crystal in symmetric Bragg geometry at 8 keV

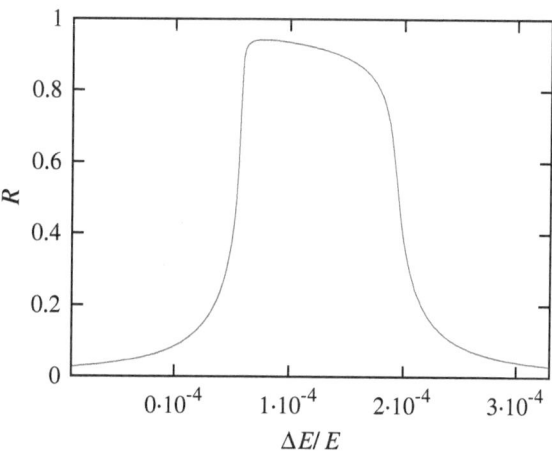

it is strongly asymmetric, being much wider in the horizontal direction. The ID10A-homepage [3] gives the size of the electron beam at the ID10-undulator as 928 μm in the horizontal direction and 23 μm in the vertical. The divergences are 24 and 9 μrad, respectively. All these values are to be understood as FWHM. Due to Als-Nielsen and McMorrow [1] the FWHM-angular divergence of the radiation cone generated by one electron is given by $\sqrt{2\lambda/L}$, where λ is the fundamental wavelength and $L = 1.6$ m the length of the undulator. This gives a value of 14 μrad at 8 keV, which conforms with the values given for the divergence of the photon beam of 28×17 μrad^2.

The distance from the source to the lenses d_1 is 33.5 m. The width of the SS0-slits can be chosen by the experimenter, common values are 300 μm both in horizontal and vertical direction. The lens system has a focus length of 11.7 m at 8 keV. The distance from the lenses to the position of the sample d_2 is 12.2 m. The widths of

[3] http://www.esrf.eu/UsersAndScience/Experiments/SoftMatter/ID10A/BeamlineDescription

the sample slits can be chosen independently in the two dimensions, they are on the order of $10\,\mu m$. The distance from the sample slits to the sample d_3 should be chosen as close as possible, as the footprint on the sample gets larger with higher distances due to diffraction, leading to smaller speckles. A distance of $0.15\,m$ seems the minimum value when using a compact furnace and having dismounted the guard slits. The distance from the sample to the detector d_4 can be chosen arbitrarily within reasonable limits.

Due to Eq. 6.1.10 the goal is to maximize the product of intensity and contrast. A few preliminary observations are immediately clear: the speckles have a characteristic angular size given by the wavelength divided by the sample slit width, whereas the angular size of the detector's pixels can be scaled arbitrarily via the sample-detector distance. Once the distance is so far that the angular pixel size is clearly smaller than the speckle size, going still farther away only decreases the count rate without increasing the coherence factor β. On the other hand, for short distances the decrease in β is compensated by the increase in the intensity. The sample slit size shows an analogue behaviour: a small slit gives a lower intensity and large speckles, but β does not profit any more once the speckles are larger than the pixels. A large slit gives a high intensity but small speckles, leading to decreasing β once the speckle size is smaller than the pixel size. Another effect of the sample slit size is the coherence of the incoming radiation: once the slit is larger than the transversal coherence length, β decreases also from this side. These two processes for the decrease of β are multiplicative. The optimal set-up would therefore be the following: close the slits (trading intensity for β) to the transversal coherence length, and then reduce the sample-detector distance, gaining intensity without sacrificing in β as the speckle size will be very big.

For a quantitative treatment Eq. 6.2.5 has to be solved. The average in Eq. 6.2.5 can be performed for a given set of parameters by Monte Carlo integration: a position and a direction of the electron and a wavelength of the generated radiation is randomly chosen, subject to the respective probability distribution functions (independent Gaussian distributions for horizontal and vertical position, horizontal and vertical divergence, and the Darwin reflectivity for the wavelength). The cone of radiation is propagated to the SS0-slits. The phases of the amplitudes within the opening of the slits are modified corresponding to their path length in Be. The amplitude after the lenses is then propagated to the sample slits. Here everything outside the opening is set to zero again, the rest is propagated to the sample. A position within the pixel is randomly chosen, giving a direction of the outgoing radiation. This then specifies $A(\vec{x})$, the amplitude with which each subvolume in the sample scatters, also accounting for the different path lengths within the material leading to different absolute values of the amplitudes. For a sample in transmission geometry normal to the incident beam the amplitude can be factorized $A(\vec{x}) = A_h(x)A_l(y)A_v(z)$, allowing the problem to be efficiently solved.

In the following I take a standard set-up, then I vary single parameters and report their effects. The standard values are a wavelength $\lambda = 1.55\,\text{Å}$ with vertical focussing, SS0-slit widths of $300\,\mu m$ in both dimensions, sample slits of $7\,\mu m$ in both dimensions, a distance from the sample slits to the sample of $17\,cm$, a sample thickness of $10\,\mu m$ with an absorption length of $12\,\mu m$, and the Andor CCD (see Sect. 5.1) at a distance of 50 cm, a scattering angle $2\theta = 25°$, and an azimuthal angle $\varphi = 45°$. This gives a coherence factor β of 4.8%.

First to the primary slits: in principle it would be beneficial to open them wider, because going to $400\,\mu m$ in both directions gives a factor of about 1.32 in intensity, whereas β drops only by a factor of 0.89. The problem is that this increases the heat load on the monochromator windows, which degrade with time even with $300\,\mu m$. The value of $300\,\mu m$ seems to be the compromise between having enough intensity and being considerate of the equipment. A minor overall increase could be achieved by closing the vertical slit by about $10\,\mu m$ and opening the horizontal slit accordingly, but the increase in intensity is nearly made up for by the decrease of β.

The question whether focussing is a good idea can be answered in the affirmative: compared to the situation without lenses, it gives about a factor of 30 in intensity while decreasing β only by about 0.67. This is only due to vertical focussing, however, the increase in intensity when enabling horizontal focussing is lost again with the decrease in β.

What is left is the width of the sample slits and the sample-detector distance. The coherence factor β as a function of those two values (assuming the same value for the sample slit in horizontal and vertical direction) is shown in Fig. 6.4. The intensity is not shown as it is in a very good approximation proportional to the area of the sample slits and strictly proportional to the inverse of the square of d_4. It can be seen that for a given detector distance there is a certain finite slit width which gives the maximum coherence factor, here it is around $6\,\mu m$. This is where diffraction at the slit sets in, closing the slit further would lead to a wider footprint of the beam on the sample and therefore smaller speckles. The slit width where this happens is obviously a function of the distance from the sample slits to the sample, which was here 17 cm. With the sample in a furnace and a beam monitor after the slits, smaller distances are hardly possible.

Blindly maximizing the product of intensity and coherence factor would lead to an extremely close detector and wide slits, giving a high intensity and a low contrast. This is not what one really wants, however. Apart from the fact that with decreasing detector distance the spread of the wave-vector transfer over the pixels of the detector increases (rendering suspicious their treatment as equivalent pixels), the evaluation as described in Sect. 5.4 relies on the fact that the charge clouds generated in the CCD do not overlap. Another point is that with small β the decay of the correlations can disappear behind spurious features of the auto-correlation function caused by, e.g., different sensitivities of the pixels. Therefore it seems wise to optimize the product of intensity and β under the constraint of an upper bound on the intensity (or equivalently a lower bound on the coherence factor). The solution to this problem as a function of the maximum intensity allowed is given by the line in Fig. 6.4.

At a detector distance of 50 cm slit sizes of $7\,\mu m$ are obviously a good choice. If the intensity is low and the coherence factor high enough, one could gain by reducing the distance further. Here the optimal slit size is already influenced by diffraction at the slits, this is corroborated by the observation that increasing the slit size in

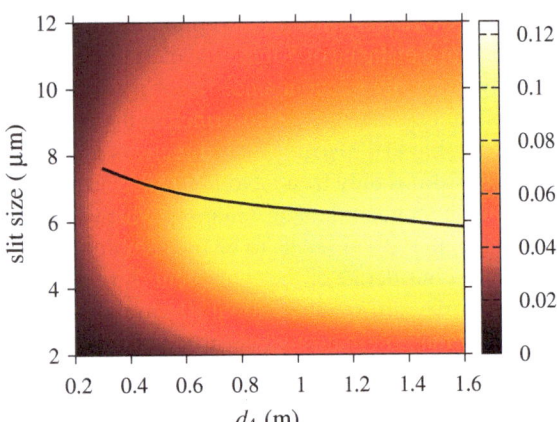

Fig. 6.4 Scan of the coherence factor β as a function of the detector distance d_4 and the width of the sample slits (set equal in horizontal and vertical direction), for the remaining values see text. The *line* shows the settings giving optimal intensity for a given β

one dimension and reducing it by the same factor in the other dimension always lowers β, a square aperture is therefore the best choice.

One should be careful in taking the results presented here at face value. First, the values of the coherence factor β obtained here are probably too high. This is because all the optical elements were considered as ideal, whereas, for instance, the energy after the monochromator will have a wider distribution in reality because of oxidation of the crystal due to the heat load. Another source of uncertainty are the distances used here: changing the distance between the lenses and the sample by 20 cm already has a discernible effect, and this clearly holds also for the distance from the source to the lenses, where the source is actually not a point but three undulators in a row, each 1 m long. The qualitative conclusions drawn here, however, should be valid.

References

1. J. Als-Nielsen, D. McMorrow, *Elements of Modern X-ray Physics* (Wiley, Chichester, 2001)
2. C.G. Darwin, The theory of X-ray reflexion. Philos. Mag. **27**, 675 (1914)
3. P. Falus, L.B. Lurio, S.G.J. Mochrie, Optimizing the signal-to-noise ratio for X-ray photon correlation spectroscopy. J. Synchrotron Rad. **13**, 253 (2006)

Chapter 7
Experimental Results

The culmination of the thesis is presented in this chapter: the results from measurements of atomic diffusion in several systems. These measurements were done during four beamtimes at ID10A at the ESRF (European synchrotron radiation facility) in Grenoble, France:

- HS-3419 from September 19th to 25th, 2007. This beamtime dealt with the coarsening of Ag precipitates in glass. We used the last two days for a feasibility test of the measurement of atomic diffusion in the metallic glass and in $Cu_{90}Au_{10}$. Our local contact was Federico Zontone.
- HD-228 from April 16th to 22nd, 2008. This beamtime was allocated for the study of the metallic glass, which we did in the first half. We used the rest for the measurements on $Cu_{90}Au_{10}$ reported in Sect. 7.1. Local contact was Andrei Fluerasu.
- HE-2845 from February 28th to March 3rd, 2009. This beamtime was awarded for studying Cu–Au. Having done that already we instead used it for remeasuring the metallic glass and anticipating the next beamtime and measuring $Si_{89}Ge_{11}$. The local contact was Anders Madsen.
- HS-3839 from April 22nd to 27th, 2009. The aim of this beamtime was to study atomic diffusion in Si–Ge. Apart from that we also did an unsuccessful feasibility test of studying diffusion in $Fe_{65}Al_{35}$. The local contact was again Federico Zontone.

For all the experiments reported here the X-ray energy was set to 8 keV (unless otherwise noted), corresponding to a wavelength of 1.55 Å, monochromatized by a Si-(111) crystal as described in Sect. 6.2.

7.1 $Cu_{90}Au_{10}$

This alloy was already considered in my diploma thesis as the most auspicious system for this kind of experiment: Cu has a relatively low absorption coefficient for X-rays of 8 keV, the preferred energy at ID10A, Au has a very high solubility in Cu at the

M. Leitner, *Studying Atomic Dynamics with Coherent X-rays*,
Springer Theses, DOI: 10.1007/978-3-642-24121-5_7,
© Springer-Verlag Berlin Heidelberg 2012

relevant temperatures, and the difference in the atomic number (and therefore electron density) between Cu and Au is among the highest possible for miscible elements. All these aspects taken together imply a high value for the diffuse scattered intensity, which is the limiting factor in today's atomic scale XPCS.

We did measurements on this system during two beamtimes: at HS-3419 in September 2007 the last day was used for a feasibility test where the amount of scattered photons with a coherent set-up was ascertained, also a measurement run at a fixed position in reciprocal space at room temperature, at 287, and 307°C, respectively, was done. The measured data showed that the amount of scattered radiation was sufficient and that at room temperature the scattered intensity showed static correlations at non-zero contrast, whereas at the elevated temperatures the contrast had vanished. This was already a result, showing that diffusion in $Cu_{90}Au_{10}$ at room temperature happens at timescales that were too long to be accessible with XPCS, whereas timescales were too short at 287°C. The principal measurements were done in April 2008 during the second half of beamtime HD-228; the results from these measurements have been published in Leitner et al. [3] and will be presented in the following.

The system was discussed in Sect. 4.3, in short: at the relevant temperatures it is a solid solution of Au in the face-centred cubic Cu crystal showing short-range order. The sample used was the same single crystal as was used by Schönfeld et al. [4], who give the actual composition as determined by X-ray fluorescence analysis to be within 0.2 at.% of the nominal value.

For the experimental set-up see Sect. A.3. We used the old furnace with the flight tube attached to it, both filled with He. The temperatures reported in the following are to be understood as the temperatures felt by the thermocouple, but due to the He atmosphere the discrepancies should be very small. We used the PI CCD with a sample-camera distance of 1.32 m. The exposure time per frame was 10 s, due to the time required for data transfer the actual frame rate was 12.34 s. We used the Be-CRL system for focussing in the vertical direction. The sample slits were set to 9 μm in the horizontal and 12 μm in the vertical direction. The count rate was about 10,000 photons per frame or 3 photons per pixel over a measurement run of typically 600 frames.

The sample was cut with a wire saw and ground to a nominal thickness of 12 μm. As we used a spot on the sample with a transmission of 0.18 at 8 keV, the actual thickness comes out as 18 μm assuming an absorption length of 10.5 μm as given by Vegard's rule.

Nominally, the sample was cut along the $(1\bar{1}0)$-plane as determined by Laue backscattering and mounted with the surface normal to the incident beam such that the [001]-direction was at an azimuthal angle of $\varphi = 45°$ and the [110]-direction at an azimuthal angle of $\varphi = -45°$ with the azimuthal angles measured with respect to the horizontal plane (for an illustration of the angles see Fig. A.2). After the experiment the sample was left in the furnace and the actual orientation was determined, according to which the orientation can be reproduced by positioning the sample in the nominal orientation, turning it around the direction of the incident beam such that the [001]-direction is at an angle of $\varphi = 51°$, then to tilt it by 1.8° so that its top goes

Fig. 7.1 Auto-correlation function for a measurement at 543 K with $\varphi = 45°$ and $2\theta = 25°$ together with an exponential fit

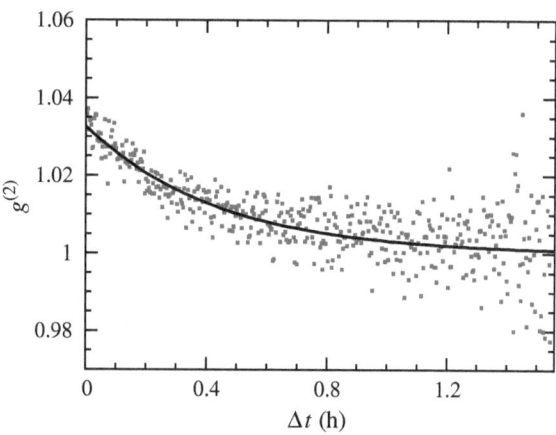

Fig. 7.2 Measured decay times as a function of the scattering angle 2θ at a temperature of 543 K and an azimuthal angle of $\varphi = 0°$ together with the values expected for nearest-neighbour jumps taking into account short-range order

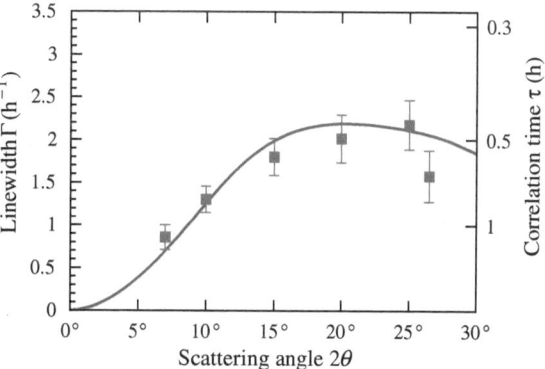

towards the beam, and finally to turn it around the vertical axis by 2.7° so that the side corresponding to $\varphi = 180°$ goes towards the beam. The actual orientation was used for the following calculations of the theoretical coherent linewidth $\Gamma_{coh}(\vec{q})$. Note that moving the detector from $\varphi = 45°$ to $\varphi = -45°$ (in the nominal orientation) for fixed 2θ corresponds to going in reciprocal space from a direction along X over L to K as defined in Fig. 4.12 (there is an equivalent point K on each of a hexagon's three sides bordering another hexagon). For a scattering angle of $2\theta = 25°$ the modulus of $\vec{q}$ is 1.75 Å^{-1}, therefore the path in reciprocal space grazes the boundary of the first Brillouin zone.

For an exemplary auto-correlation function see Fig. 7.1. Fitting was done with an exponential decay with $\vec{q}$-dependent correlation times. The fitted correlation times for several values of 2θ for fixed $\varphi = 0°$ at a temperature of 543 K is shown in Fig. 7.2. A scan of φ for the same temperature and scattering angles of $2\theta = 20°$ and $2\theta = 25°$ are shown in Fig. 7.3. The theoretical values were calculated according to Eqs. 2.3.21 and 2.3.14 for nearest-neighbour exchanges in the face-centred cubic lattice, where the values for I_{SRO} were computed from the occupations of the first 13

Fig. 7.3 Measured decay times as a function of the azimuthal angle φ at a temperature of 543 K and a scattering angle of $2\theta = 20°$ (*upper panel*) and $2\theta = 25°$ (*lower panel*) together with the values expected for nearest-neighbour jumps taking into account short-range order

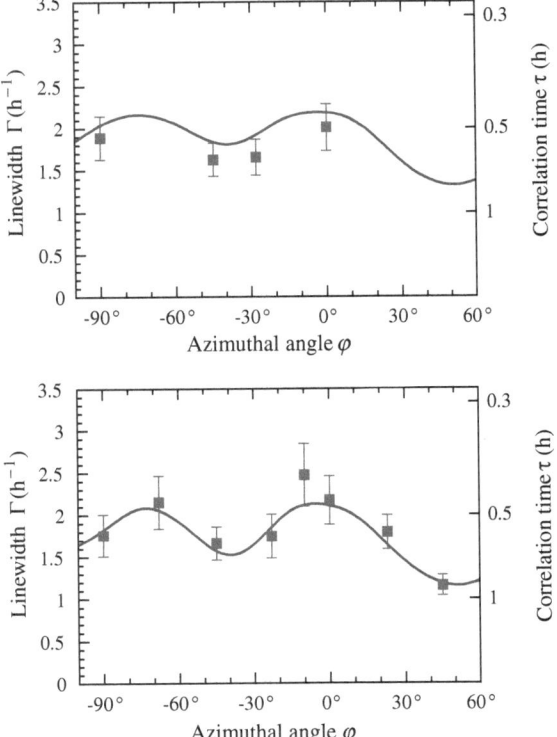

nearest-neighbour shells as given by Schönfeld et al. [4]. The only free parameter left in this theory is the raw jump frequency $\tilde{\nu}$ in the nomenclature of Sect. 2.3, which is (2230 ± 60) s at 543 K. Given the fact that these 16 independent data points were fitted by just one parameter the agreement is striking. By comparing the two panels of Fig. 7.3 one can get the impression, however, that perhaps the linewidths for $2\theta = 20°$ are too narrow in relation to the ones for $2\theta = 25°$. Also the data in Fig. 7.2 seem to hint into the direction that the increase with 2θ should be shallower around 20°. Two possible reasons can be given for this effect: On the one hand it could be due to the action of the vacancy as treated in Sect. 3.5, leading to modified effective translation vectors. Contrary to self diffusion, where the effects can be readily computed without any additional parameters [5], here the affinity of the vacancy to the constituent atoms would have to be considered, which was not attempted. The other possibility is that the temperature is already so low that deviations from the theory as treated in Sect. 4.3 appear. Note that the lower panel of Fig. 4.12. corresponds to a temperature of 530 K, which is only slightly lower than the 543 K of the experiment.

After establishing the diffusion mechanism the chemical diffusion constant as defined in Sect. 2.3 can be inferred from measuring the correlation time at any given wave-vector. This was done at $2\theta = 25°$ and $\varphi = 0°$ for a number of temperatures,

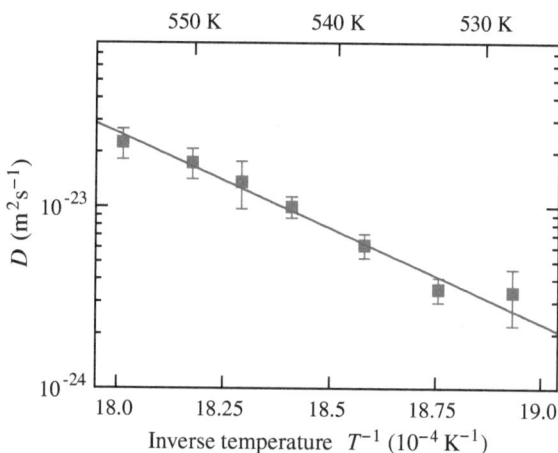

Fig. 7.4 Chemical diffusion constant as a function of temperature with an Arrhenius fit

the results of which are given in Fig. 7.4. As shown in Sect. 6.1 the relative errors due to counting noise get larger both for long and short correlation times. This does not seem to be a problem here, but it is quite possible that the value corresponding to the lowest temperature is already affected by instabilities of the beam.

The variation of the chemical diffusion coefficient with temperature is due to the variation of the jump frequency $\tilde{\nu}$. There are two contributions to this variations: on the one hand it is E_S, the mean energy necessary to raise the jumping atom to the saddle point on the energy landscape, on the other hand also the vacancy concentration is thermally activated. For the small range of temperatures treated here all the other aspects can be considered as temperature-independent, therefore it is valid to fit the diffusion coefficients by an Arrhenius dependence. This was done in Fig. 7.4, giving an activation enthalpy of $E_A = (2.09 \pm 0.15)$eV. There are no measurements of chemical diffusion in Cu–Au to compare this value with, but as the Au-atoms repel each other as nearest neighbours, chemical diffusion has to be closely linked to tracer diffusion of Au in Cu$_{90}$Au$_{10}$. The activation enthalpy obtained here is in very good accordance with the values of 2.0–2.2 eV obtained for the tracer diffusion of Au in Cu [1].

7.2 Fe$_{65}$Al$_{35}$

This system was measured only briefly at the beamtime HS-3839. Iron is a problematic constituent for such experiments as will become clear below, we considered it only because a single crystalline sample was already available from preceding Mößbauer experiments.

Just as NiAl, FeAl displays the B2-ordering. The surplus Fe-atoms are incorporated as structural antisites on the Al-sublattice. We would therefore expect diffuse

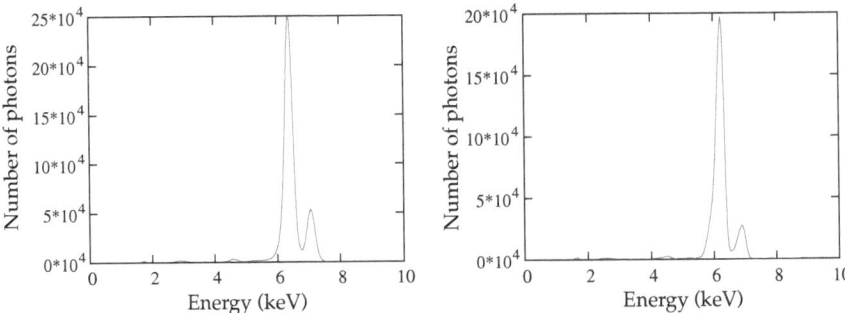

Fig. 7.5 Histograms of detected photon energies for an incident photon energy of 7.1 keV (*left*) and 8.0 keV (*right*)

scattering from these Fe-antisites. Iron has its K-edge at 7.112 keV, hence we set the undulator to an energy of 7.1 keV. The necessity for going below the edge is only in part due to the increased absorption above the edge, the more important point is to avoid fluorescence which would drown the elastic diffuse intensity.

The left panel of Fig. 7.5 shows a histogram of the photon energies detected with an incident energy of 7.1 keV, using the reference value of 1955 ADUs corresponding to an 8 keV-photon as established in Sect. 5.3. It seems that there are elastically scattered photons (photons with an energy of 7.1 keV), but there is obviously also FeKα-fluorescence at 6.4 keV. Evaluating only the apparently elastically scattered photons showed no contrast, however, even when the sample was at room temperature. Also the scattered intensity was extremely low at 350 counts per frame of 20 s.

We also tried to measure with an incident energy of 8.0 keV. The histogram from this experiment is shown in the right panel of Fig. 7.5. While the histogram from the measurement with 7.1 keV corresponded to an accumulated exposure of nearly 7 h, here it is an accumulated exposure of only 10 s. The overall scattered intensity is therefore higher by a factor of about 2,500, while the incident intensity is only higher by a factor of 5. Therefore the scattered photons at 8.0 keV were nearly exclusively due to fluorescence, also their energy distribution fits with FeKα and FeKβ. Note that the apparent similarity of the two panels of Fig. 7.5 suggests that also with the nominal incident energy of 7.1 keV most of the "elastic" photons were actually FeKβ-fluorescence.

What can be learned from this experiment is that the energy distribution after the monochromator as given by Fig. 6.3 is only the ideal case. In reality the inevitable degradation due to the high heat load will lead to more pronounced tails of the reflectivity curve, it is therefore highly recommendable to have a wider safety margin between the incident energy and the absorption edge. The reduction by a factor of 5 in the incident intensity when going from 8.0 to 7.1 keV is mostly due to the change in the focal length of the Be-CRL system. This factor could therefore be gained by using the ID10C-branch (which can focus at 7 keV), but still the intensity would probably be too low for measurements at the ESRF.

7.3 Si$_{89}$Ge$_{11}$

Extensive experiments on this system were done during the beamtimes HE-2845 and HS-3839 in March and April of 2009 using both furnaces (see Sect. A.3). It was selected for its differences from metallic systems and because a successful experiment would provide a definitive answer as to which mechanism of diffusion (from among the number of possible candidates as given in Sect. 4.1) is at work in this system, a question which is still open.

One main distinction from the other systems treated here is the high activation enthalpy of diffusion. This is a significant difference as became clear in the course of the experiments. What we observed in the two-time auto-correlation functions were apparent dynamics that got slower with time, prompting us to raise the temperature again and again. Such a behaviour could in principle result from kinetics-driven dynamics, this means that the system is out of equilibrium after changing the temperature, causing the atoms to rearrange for the new equilibrium. This explanation seemed highly suspicious as there should be no short-range order in SiGe at such elevated temperatures. After chasing these "dynamics" from 500 to 870°C, with the decay of the correlations being fast each time after raising the temperature and then slowly dying away, we were sure that something was wrong, because if the equilibrium dynamics were too slow to be measurable at 870°C, any atomic motion at 500°C would be unthinkable. Moreover, sometimes the two-time auto-correlation function showed artefacts as in Fig. 7.6. Computing the ordinary auto-correlation function for seemingly undisturbed stretches such as the first 290 frames in Fig. 7.6 yielded a form as given in Fig. 7.7, which could satisfactorily be fitted with auto-correlation functions of the form $1 + \beta \exp(-2(\Delta t/\tau)^2)$ with apparently $\vec{q}$-dependent correlation times.

At first I thought that this behaviour was due to the sample moving slowly with respect to the beam, leading to the gradual loss of the correlations on the timescale of a given point in the sample transversing the illuminated area. It seemed unrealistic, however, that this velocity of 10 μm per half hour and sometimes even faster could have been maintained for days. Also this would lead to a linear decay with time of the correlations for small times. The true reason for this puzzling behaviour became clear to me only recently: a quick calculation shows that rotating the sample by just 0.002° moves the speckle pattern in reciprocal space by a speckle width. Assuming a speckle form of $(\sin(x)/x)^2$ and folding with the auto-correlation of the characteristic function of the pixel gives a shape very similar to a Gaussian distribution, which explains the observed form of the intensity auto-correlation function. This also explains the fact that with higher scattering angles the correlations decay faster. Figure 7.6 can now be understood: the sample is subject to a steady rotation, at frame 290 it jumps back to the position corresponding to frame 180, and it proceeds with the original movement, leading to off-diagonal correlations. It moves now a bit faster so that already at frame 370 it passes through the position it had immediately before the jump.

Fig. 7.6 Two-time auto-correlation function of 434 frames of a measurement of $Si_{89}Ge_{11}$ at 840°C in pseudo-color display, smoothed by a Gaussian kernel with a width of three frames. The singular event is at frame 290

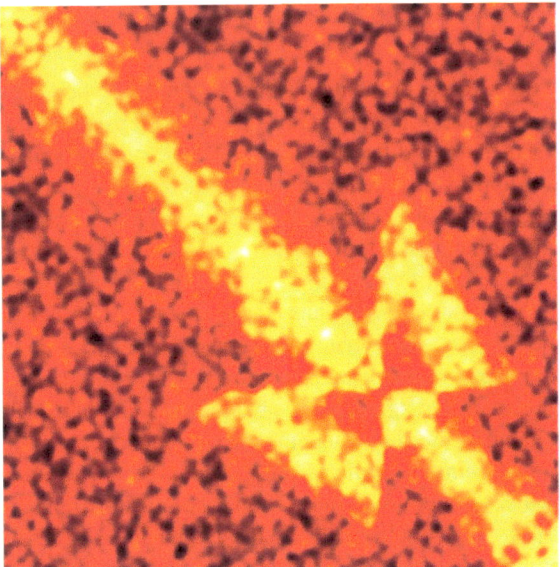

Fig. 7.7 Auto-correlation function for a measurement at 800°C together with a fit of the form $1 + \beta \exp(-2(\Delta t/\tau)^2)$

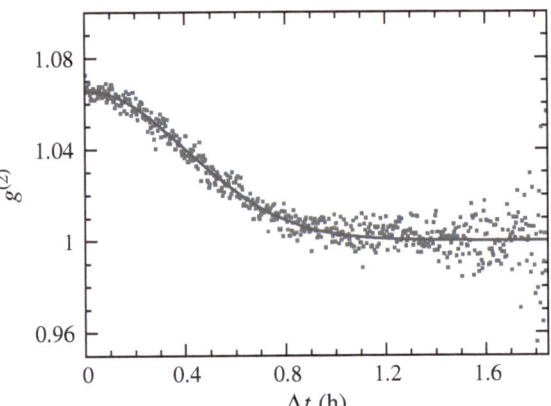

This rotating of the sample is obviously due to temperature as changing the temperature leads to an initially faster loss of correlation. It cannot be caused by solely elastic expansion and contraction, however, because the motion is still there even hours after changing the temperature. It is also not due to fluctuations in the temperature; these were on the order of a few tenths of a degree, whereas one has to change the temperature by five degrees during the measurement to cause a correlation loss. Probably it comes from the plastic release of stresses in some parts of the furnace, or the ill-defined situation produced by clamping the brittle sample into the sample holder and necessarily crumbling it partially.

A concluding remark should be that the temperatures of up to 870°C used here seem very high compared to literature data [2], all the more given the fact that these temperatures were seemingly too low for diffusion to happen. As explained in more detail in Sect. A.3, the temperatures quoted here are the temperatures of the sample holder (felt by the thermocouple), which glowed much brighter than the sample at such elevated temperatures. We have yet to ascertain the relation between the temperature of the sample holder and the sample, but definitely a diffusion experiment in SiGe needs furnaces allowing still higher temperatures, which is quite a problem due to the need for a large solid angle for the exiting radiation.

7.4 The Metallic Glass $Zr_{65}Cu_{17.5}Ni_{10}Al_{7.5}$

This system is a so-called bulk metallic glass, that means during cooling from the liquid to the solid state moderate cooling rates suffice to prevent crystallization, allowing bulk samples to be prepared in a glassy state. This system therefore constitutes an opportunity to extend our studies to diffusion in non-crystalline media, a field where many questions are still left to be answered (see Sect. 4.4). A more practical reason for studying this system is that XPCS deals with diffuse scattering, of which there is much available here.

Apart from the feasibility test during the beamtime HS-3419 we did the main measurements on this system during the beamtimes HD-228 and HE-2845. For the experimental set-up see Sect. A.3. During the second beamtime we had the same problems as described in Sect. 7.3, we thought that we measured dynamics while in reality it was most probably the sample moving. These measurements were done between 300 and 327°C and for different scattering angles 2θ. In beamtime HD-228, however, we paid more attention to covering a wide range of temperatures. We performed measurements for temperatures from 260 to 370°C using the old furnace. This wide range of temperatures led to partial crystallization as can be seen from the small Bragg peak at about 31.5° in Fig. 7.8. These measurements were therefore done under ill-defined conditions. Also we witnessed ageing, i.e. revisiting a temperature yielded slower dynamics than in the first measurement, but this contributes to my confidence that what I report in the following are really the dynamics of the sample and not of the sample mount.

We did our experiment in the short-range order peak of the glass at $2\theta = 37°$. The two-time auto-correlation functions of the measurements at 300°C showed correlation losses with a frequency of about two per hour, this means that there were singular events where obviously something moved, destroying the correlations between the frames before and after this event. The correlation times due to the dynamics in the sample were seemingly on longer timescales. For higher temperatures such correlation losses are not visible in the two-time auto-correlation functions. This is probably due to the fact that the sample correlation times are shorter, so there are hardly correlations over longer timescales that could be destroyed by these singular events. Assuming that they happen about as frequently as at lower temperatures implies that

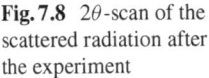

Fig. 7.8 2θ-scan of the scattered radiation after the experiment

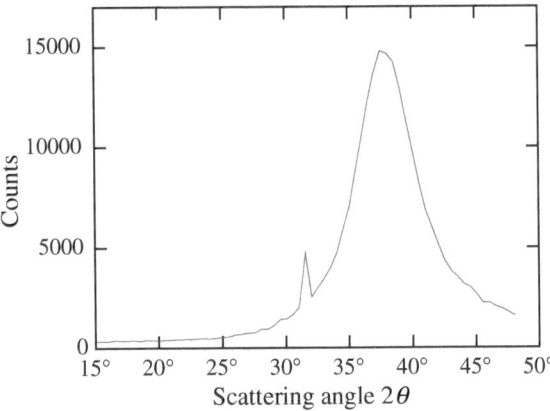

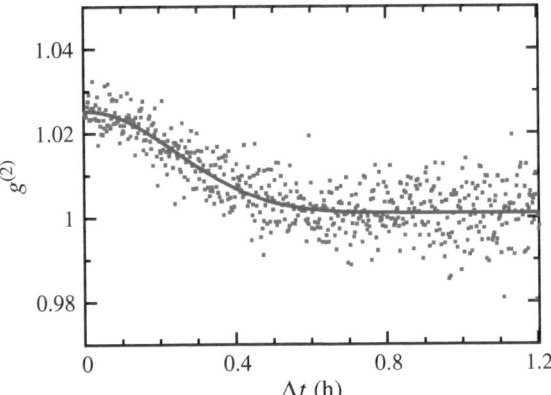

Fig. 7.9 Auto-correlation function for a measurement at 330°C together with a fit of the form $1 + \beta \exp(-2(\Delta t/\tau)^\gamma)$

the auto-correlation function is only weakly affected by them. This is corroborated by the fact that the fitted correlation times show a strong temperature dependence.

In non-crystalline media it is usually found that the correlation functions cannot be fitted with a simple exponential decay. By convention one then introduces an additional parameter γ and fits the expression $1 + \beta \exp(-2(\Delta t/\tau)^\gamma)$ to the data, which gives always a good fit from the phenomenological viewpoint. Here γ was always around 2, giving compressed exponential decays as illustrated in Fig. 7.9. For the fitted correlation times as a function of the temperature see Fig. 7.10. The estimated errors of the fitted τ are here a few percent, which is about the point size. Especially at lower temperatures the measured correlation times will likely be too short because of the possibility of hidden correlation losses. I want to restate that these are non-equilibrium values, they depend on the sample history. Here the sample was kept for about two hours at 370°C prior to these measurements, then the measurements were done in the order of ascending temperatures, with the sample being kept at each temperature for about two hours. We intended these measurements only for qualitative information, therefore the correlation times given in Fig. 7.10 are

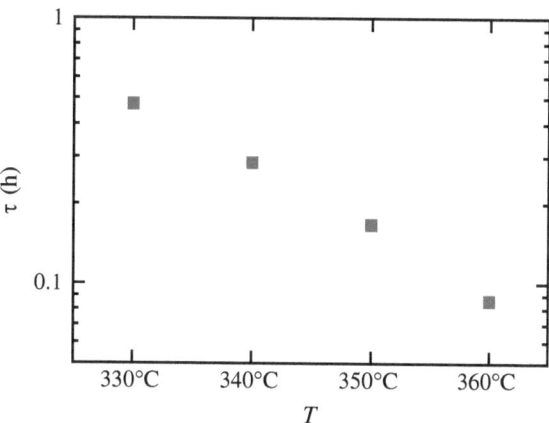

Fig. 7.10 Correlation times for different temperatures

probably not well reproducible. This is the reason why I refrain from fitting an activation enthalpy.

For a further experiment on this system one can use the information that the interesting temperature range is around 350°C. Keeping the sample at this temperature for a long time before the experiment should lead to a so-called quasi-equilibrium, which means that although the sample is not in equilibrium (which can never be the case for a glass), the kinetic relaxations happen on much longer timescales than the experiment. In this manner one could record the variation of the correlation time τ and the compressing parameter γ with the scattering angle 2θ, allowing to gain insight into the dynamics at work on the atomic scale.

References

1. S. Fujikawa, M. Werner, H. Mehrer, A. Seeger, Diffusion of gold in copper over a wide range of temperature. Mater. Sci. Forum **15**, 431–436 (1986)
2. R. Kube, H. Bracht, J. Lundsgaard Hansen, A. Nylandsted Larsen, E.E. Haller, S. Paul, W. Lerch, Simultaneous diffusion of Si and Ge in isotopically controlled $Si_{1-x}Ge_x$ heterostructures. Mat. Sci. Semicond. Proc. **11**, 378 (2008)
3. M. Leitner, B. Sepiol, L.M. Stadler, B. Pfau, G. Vogl, Atomic diffusion studied with coherent X-rays. Nat. Mat. **8**, 717 (2009)
4. B. Schönfeld, M.J. Portmann, S.Y. Yu, G. Kostorz, The type of order in Cu-10 at.% Au—evidence from the diffuse scattering of X-rays. Acta Mater. **47**, 1413 (1999)
5. C.A. Sholl, Diffusion correlation factors and atomic displacements for the vacancy mechanism. J. Phys. C **14**, 2723 (1981)

Chapter 8
Outlook

Here I want to give some concluding remarks and present an outlook about what I think would be rewarding directions of future research.

The main point of this thesis was the first successful demonstration of applying XPCS to study dynamics on the fundamental spatial scale of condensed matter physics, the atomic scale. X-rays with their wavelengths on the order of atomic distances are naturally suited for this task, therefore soon after the first demonstrations of XPCS the idea arose to apply it to measure atomic diffusion. The necessary intensity has become available only recently, though. The other possible direction to go in using coherent X-rays for studying dynamics is to smaller timescales. These two ambitions, smaller length scales or smaller timescales, exclude each other with today's sources, because both need more intensity; the experiments shown here are limited to correlation times of minutes or longer. For atomic diffusion this is not a big issue, because by choosing the temperature appropriately the dynamics in the sample can generally be made to happen on accessible timescales, and the dynamics at a hundred degrees more are very likely qualitatively the same. With the availability of the X-ray free electron lasers, however, the fundamental scale (pico- to femtoseconds) should become accessible also in the time domain. Contrary to the experiments presented here, where the atomic positions before and after the jump are compared, this would enable us to follow the atoms *during* their jumping. For me this seems to be the ultimate goal in our branch of solid state physics.

In contrast to the qualitatively new kinds of experiments that the X-ray free electron lasers will bring, the new synchrotrons of the third generation like PETRA III will enable us to do experiments such as those presented here, but without being restricted to strongly scattering systems as is now the case. Such experiments will probably never become a standard technique of sample characterization, but I think that they will become the method of choice for studying the mechanisms of atomic motion in a few representative systems, allowing us to develop a comprehensive picture of what happens at the atomic scale in solids.

Now to the concrete points: I think that it would be very rewarding to make a thorough investigation of the correlation times as a function of the wave-vector

M. Leitner, *Studying Atomic Dynamics with Coherent X-rays*,
Springer Theses, DOI: 10.1007/978-3-642-24121-5_8,
© Springer-Verlag Berlin Heidelberg 2012

in Cu–Au. This could be achieved by having a more stable set-up, so that one can be certain that the measured correlation times are only due to the sample, or by having more intensity, allowing to measure shorter correlation times (which are more robust against instabilities). In particular a measurement of a sample with a slightly higher Au content, say $Cu_{85}Au_{15}$, would be interesting. There one could probably unambiguously detect deviations from the first-order theory; as I have demonstrated in Sect. 4.3 this would give information on the influence of the jumping atom's neighbourhood on the barrier height it has to overcome. Such data have never been obtained yet. One could compare them to the results from ab-initio calculations, which would be to my knowledge the first possibility to test predictions for states far from the relaxed ground state.

The second point I want to emphasize is diffusion in Si–Ge. Even though diffusion in semiconductors is such an important topic (both for production and thermal stability), the question of how this happens is still not settled. We need to solve the problem of the stability of the sample in the furnace at high temperatures first, but then it would be very easy to decide on the diffusion mechanism.

Appendix

A.1 Eigenvalues of $\mathbf{K}'$

Here I prove that the matrix $\mathbf{K}'(q)$ as defined in Sect. 2.2 has non-positive eigenvalues, a fact which is related to Gerschgorin's circle theorem [1]. Actually I prove it for the matrix $\hat{\mathbf{K}}(q)$, which has the same eigenvalues as $\mathbf{K}'(q)$ because the two matrices are similar due to Eq. 2.2.9.

Let λ be an eigenvalue of $\hat{\mathbf{K}}(q)$ with eigenvector e, that is

$$\sum_{\mu} \left(\hat{\mathbf{K}}(q) \right)_{\nu,\mu} e_{\mu} = \lambda e_{\nu} \quad \forall \nu. \tag{A.1.1}$$

Now take a fixed ν such that

$$|e_{\nu}| \geq |e_{\mu}| \quad \forall \mu. \tag{A.1.2}$$

So Eq. A.1.1 can be restated as

$$\begin{aligned}
\lambda &= \left(\hat{\mathbf{K}}(q) \right)_{\nu,\nu} + \sum_{\mu \neq \nu} \left(\hat{\mathbf{K}}(q) \right)_{\nu,\mu} \frac{e_{\mu}}{e_{\nu}} \\
&= \sum_{\Delta x} \left(\mathbf{K}(\Delta x) \right)_{\nu,\nu} e^{-iq\Delta x} + \sum_{\Delta x} \sum_{\mu \neq \nu} \left(\mathbf{K}(\Delta x) \right)_{\nu,\mu} \frac{e_{\mu}}{e_{\nu}} e^{-iq\Delta x} \\
&= \left(\mathbf{K}(o) \right)_{\nu,\nu} + \sum_{(\Delta x,\mu) \neq (o,\nu)} \left(\mathbf{K}(\Delta x) \right)_{\nu,\mu} \frac{e_{\mu}}{e_{\nu}} e^{-iq\Delta x}.
\end{aligned} \tag{A.1.3}$$

M. Leitner, *Studying Atomic Dynamics with Coherent X-rays*,
Springer Theses, DOI: 10.1007/978-3-642-24121-5,
© Springer-Verlag Berlin Heidelberg 2012

Taking the real part shows that

$$
\begin{aligned}
\mathrm{Re}\,(\lambda) &= \left(\mathbf{K}(\mathbf{o})\right)_{v,v} + \mathrm{Re}\left(\sum_{(\Delta x,\mu)\neq(\mathbf{o},v)} \left(\mathbf{K}(\Delta x)\right)_{v,\mu} \frac{e_\mu}{e_v} e^{-iq\Delta x}\right) \\
&\leq \left(\mathbf{K}(\mathbf{o})\right)_{v,v} + \left|\sum_{(\Delta x,\mu)\neq(\mathbf{o},v)} \left(\mathbf{K}(\Delta x)\right)_{v,\mu} \frac{e_\mu}{e_v} e^{-iq\Delta x}\right| \\
&\leq \left(\mathbf{K}(\mathbf{o})\right)_{v,v} + \sum_{(\Delta x,\mu)\neq(\mathbf{o},v)} \left|\left(\mathbf{K}(\Delta x)\right)_{v,\mu}\right| \left|\frac{e_\mu}{e_v}\right| \left|e^{-iq\Delta x}\right| \\
&\leq \left(\mathbf{K}(\mathbf{o})\right)_{v,v} + \sum_{(\Delta x,\mu)\neq(\mathbf{o},v)} \left(\mathbf{K}(\Delta x)\right)_{v,\mu} = 0, \qquad (\mathrm{A.1.4})
\end{aligned}
$$

where the last equality is due to Eq. 2.2.1.

Therefore also the real parts of the eigenvalues of $\mathbf{K}'(q)$ are less or equal to zero, and as $\mathbf{K}'(q)$ is Hermitian, the eigenvalues are real.

A.2 The Interplay of Fluctuation and Relaxation

Here I give the proof of Eq. 2.3.20. For the nomenclature I refer the reader to Sect. 2.3. Let the system consist of N lattice sites, of which cN are occupied. Take a wave-vector q and a state of the system σ. The quantity of interest is the amplitude A, the Fourier transform of σ. I assume that the amplitude is on the order of $\sqrt{N}$, this assumption will be justified later.

The assumption that the amplitude is on the order of $\sqrt{N}$ implies that the contributions of the distinct particles to the overall amplitude are in first order uniformly distributed on the complex unit circle, the relative deviations from the uniform distribution are only on the order of $1/\sqrt{N}$. Therefore the variance of the real part of the overall amplitude after exactly one particle has made a jump is given by

$$
\begin{aligned}
v_r &= \left\langle \left(\cos(q(x+\Delta x)) - \cos(qx)\right)^2\right\rangle + O(1/\sqrt{N}) \\
&= \left\langle \cos^2(q(x+\Delta x)) - 2\cos(q(x+\Delta x))\cos(qx) + \cos^2(qx)\right\rangle \\
&= \left(\frac{1}{2} - \left\langle \cos(2qx + q\Delta x) + \cos(q\Delta x)\right\rangle + \frac{1}{2}\right) = \left\langle 1 - \cos(q\Delta x)\right\rangle \quad (\mathrm{A.2.1})
\end{aligned}
$$

in the leading order of N, analogously for the imaginary part v_i. The exchange rate v for a given site is

$$
v = \sum_{\Delta x} \tilde{v}_{\Delta x}, \qquad (\mathrm{A.2.2})
$$

leading to an increase rate of the variance of the overall amplitude's real part

$$\frac{d}{dt}V(A(\boldsymbol{q})) = Nc(1-c)vv_r = Nc(1-c)\sum_{\Delta x}\tilde{v}_{\Delta x}(1-\cos(\boldsymbol{q}\Delta\boldsymbol{x}))$$
$$= Nc(1-c)\Gamma_{\text{inc}}(\boldsymbol{q}). \tag{A.2.3}$$

The same holds for the imaginary part. The factor $c(1-c)$ is the first-order approximation for the exchanges leading to a change in the amplitude. The behaviour of the expected value of the amplitude was already given in Eq. 2.3.13:

$$\frac{d}{dt}\langle A(\boldsymbol{q})\rangle = \langle A(\boldsymbol{q})\rangle\Gamma_{\text{coh}}(\boldsymbol{q}) \tag{A.2.4}$$

Picture now an ensemble of systems, i.e. a distribution of amplitudes in the complex plane. Equation A.2.3 acts as a convolution with a normal distribution with infinitesimal width, whereas Eq. A.2.4 acts as an infinitesimal contraction. Due to the interplay between these two processes the distribution of amplitudes will evolve to a normal distribution, whose variance (squared width) of the real component V_r can be computed by requiring stationarity:

$$0 = \frac{d}{dt}V_r = Nc(1-c)\Gamma_{\text{inc}}(\boldsymbol{q}) - 2V_r\Gamma_{\text{coh}}(\boldsymbol{q}), \tag{A.2.5}$$

the same holds for the variance of the imaginary component V_i. The expected value of the intensity is the expected value of the modulus of the squared amplitude, that is

$$I_{\text{SRO}}(\boldsymbol{q}) = V_r + V_i = \frac{Nc(1-c)\Gamma_{\text{inc}}(\boldsymbol{q})}{\Gamma_{\text{coh}}(\boldsymbol{q})} = \frac{1}{\left(1 + \frac{\hat{V}(q)c(1-c)}{kT}\right)}, \tag{A.2.6}$$

proving Eq. 2.3.20.

A.3 The Sample Environment

Here I give a brief description of the rest of the experimental apparatus, i.e. what is between the beamline optics (see Sect. 6.2) and the CCD-camera (Chap. 5).

A fundamental distinction in scattering experiments concerns the position of the sample with respect to the optical path. This can either be the so-called transmission geometry, where the sample is thin (on the order of the X-ray absorption length), the primary beam passes through the sample, and the scattered radiation is detected at the downstream side of the sample. The other possibility is scattering geometry, where the scattered radiation is detected on the side of the sample that faces the primary beam. Both possibilities have their assets and drawbacks: for transmission geometry the preparation and the handling of the thin sample can be a problem, especially under the aspect that it should be a

Fig. A.1 The heart of the old
furnace. The sample holder is
mounted inside, the heating
wire is coiled outside

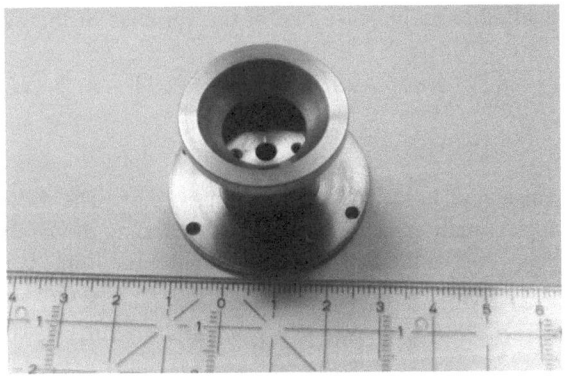

single crystal. With scattering geometry one can use a comfortably thick sample, and it is principally conceivable to use a powder sample.[1] Also there is no exiting beam in scattering geometry, whereas in transmission geometry one has to intercept the transmitted beam before it hits something and contributes to the background via elastic scattering or fluorescence. The advantages of the transmission geometry are first the accessibility of the whole azimuthal angle without having to rotate the sample and second the robustness to thermal fluctuations: in a furnace with rotational symmetry with respect to the incident beam thermal fluctuations have in first order no influence on the position of the sample. In scattering geometry, however, thermal expansion will move the sample normal to the beam, destroying the correlations in the scattered radiation. For these reasons up to now all our experiments were done in transmission geometry.

Apart from the first feasibility test during the beamtime HS-3419, where a furnace for small-angle scattering was used, the experiments presented in this thesis used two furnaces specially designed for wide-angle scattering in transmission geometry. At the heart of both is a drum (see Fig. A.1) around which a wire is coiled, used for resistive heating. They differ in the fact that the old one uses an electrically insulated wire, whereas the new one uses a non-insulated wire from a high-resistivity alloy, insulated by the high-temperature adhesive into which it is embedded. This should allow higher temperatures. Inside this drum the sample holder is mounted, which is compatible to both furnaces. The drums are designed in order to cover as much as possible of the solid angle while still allowing scattering angles of $2\theta = 40°$. The drums have a bore where a thermocouple is inserted. The relation between the thermocouple's temperature and the temperature of the sample is still an open question for us, although

---

[1] This can be an option if a single crystal cannot be grown. A powder of the sample material with grains on the order of 100 nm interspersed with an X-ray-transparent substance such as boron nitride would give a well-defined sample as opposed to a polycrystal which will recrystallize under elevated temperatures and where the beam hits a small number of crystallites of unknown orientation.

Fig. A.2 The set-up with the old furnace connected to the flight tube. Also visible is the PI CCD

experience gained on similar furnaces tells us that at temperatures around 300°C the discrepancy should not be more than 5°C. At temperatures around 800°C, however, where the drum and the sample holder can be seen, they are much brighter than the sample. This is because the sample is heated via heat conduction (which scales with the temperature difference), but it loses its heat via radiation (which scales with the fourth power of the temperature difference). Covering part of the opening angle with Ta heat shields and wrapping the drum in Al foil helps in reaching high temperatures, although the Kapton windows then slowly become fogged by the evaporating Al.

The temperature was regulated by a controller via the voltage applied to the heating wire. This is done via a set of three parameters; depending on how good these parameters were set the temperature showed fluctuations around the nominal value of 0.1−1°C. This was no problem, however, as we had experimentally ascertained that the correlations in the scattered radiation were lost only with temperature differences of 5° or more.

For the set-up see Fig. A.2. The furnaces can be connected directly to the flight tube via flexible bellows, in fact there exists an adapter so that it can be connected to two flight tubes. This possibility was used only for the measurements of the metallic glass during beamtime HD-228, because it proved to be quite demanding to evaluate two data sets in parallel and to keep two cameras busy in a purposeful way. Anyway, the availability of two working CCDs is only rarely given. Also the possibility for connecting one flight tube directly to the furnace was used only in the other half of beamtime HD-228 for the measurements on $Cu_{90}Au_{10}$ reported in Sect. 7.1. For being able to move the flight tube it turned out to be necessary to fill the furnace and the attached flight tube with He at ambient pressure because of the forces the atmospheric pressure would exert onto evacuated bellows. At the other experiments the furnace and the flight tube were evacuated. Kapton foil was used for the furnace and flight tube windows.

Our single crystal samples were mounted with the surface normal to the incident beam and they were not moved during the experiments. With the samples made of metallic glass orientation was no issue, therefore the orientation of the furnace (including the sample) was adapted in order to more easily reach high scattering angles.

The furnace is mounted on a stage which can be rotated (and also translated by small distances), the flight tube and the camera are mounted on an arm which can be rotated in the horizontal plane independently from the furnace. We realized detector positions out of the horizontal plane by an improvisational construction from X95 rails, the in-plane dimension was set by moving the goniometer arm.

Reference

1. S. Gerschgorin, Über die Abgrenzung der Eigenwerte einer Matrix. Izv. Akad. Nauk SSSR **7**, 749 (1931)

Curriculum Vitae

Personal data

Name:	Michael Leitner
Date of birth:	October 13th, 1982
Place of birth:	Mistelbach, Austria
Nationality:	Austria
Marital status:	Married

Education

1989–1993:	Volksschule Neudorf
1993–2001:	Bundesgymnasium u. Bundesrealgymnasium Laa a.d. Thaya
June 2001:	Matura (school leaving examination)
2001–2002:	Military service
2002 to present:	Undergraduate studies (mathematics) at the Faculty of Mathematics, University of Vienna
2003–2007:	Undergraduate studies (physics) at the Faculty of Physics, University of Vienna

M. Leitner, *Studying Atomic Dynamics with Coherent X-rays*,
Springer Theses, DOI: 10.1007/978-3-642-24121-5,
© Springer-Verlag Berlin Heidelberg 2012

March 2007:   Mag. rer. nat. (MSc) in physics, University of Vienna Diploma
                         thesis: "Diffusion studies with X-ray photon correlation
                         spectroscopy", Advisor: ao. Univ.-Prof. Dr. Bogdan Sepiol

2007–2010:   Doctoral studies (physics) at the Faculty of Physics,
                         University of Vienna

Positions

March 2007   Research Assistent supported by the Initiativkolleg I022-N
to Dec. 2008:   "Experimental Materials Science—Nanostructured Materi-
                         als", research group "Dynamics of Condensed Systems",
                         Faculty of Physics, University of Vienna

January 2009   Research Assistent at the research group "Dynamics of
to present: Condensed Systems", Faculty of Physics, University of
                         Vienna